Andrew S. Fuller

Practical Forestry

Andrew S. Fuller

Practical Forestry

ISBN/EAN: 9783744692540

Printed in Europe, USA, Canada, Australia, Japan

Cover: Foto ©Andreas Hilbeck / pixelio.de

More available books at **www.hansebooks.com**

PRACTICAL FORESTRY.

A TREATISE ON THE PROPAGATION, PLANTING, AND CULTIVA-
TION, WITH A DESCRIPTION, AND THE BOTANICAL
AND POPULAR NAMES OF ALL THE

Indigenous Trees of the United States,

BOTH EVERGREEN AND DECIDUOUS, TOGETHER WITH NOTES ON
A LARGE NUMBER OF THE MOST

VALUABLE EXOTIC SPECIES.

BY

ANDREW S. FULLER,

AUTHOR OF THE "STRAWBERRY CULTURIST," "GRAPE CULTURIST," "SMALL
FRUIT CULTURIST," ETC., ETC.

ILLUSTRATED.

NEW YORK:
ORANGE JUDD COMPANY,
52 & 54 LAFAYETTE PLACE.
1908

CONTENTS.

———◆◇◆———

CHAPTER I.

Influence of Forests on Climate.................................... 9
Forests and Streams... 15
Trees for Shelter... 17
Forests and Insects... 18

CHAPTER II.

The Characteristics of Trees...................................... 19
The Movement of Sap in Trees...................................... 22
The Buds of Trees... 23

CHAPTER III.

Raising Trees from Seed.. 25
Preparing a Seed-bed.. 28

CHAPTER IV.

Transplanting Seedlings... 32

CHAPTER V.

Budding and Grafting.. 36
Budding.. 36
Grafting Deciduous Trees 42

CHAPTER VI.

Grafting Conifers... 48

CHAPTER VII.

Coniferæ from Cuttings.. 51
Layers.. 54

CHAPTER VIII.

Deciduous Trees from Cuttings..................................... 54
Propagation by Layering... 55

CHAPTER IX.

Seedlings of Coniferæ... 58
Sowing the Seeds.. 60
Evergreens from the Forests....................................... 64
Season for Transplanting.. 66

(3)

CHAPTER X.

Pruning Forest Trees.. 67
Time to Prune.. 69
Pruning Evergreens... 70
Implements used in Pruning... 71

CHAPTER XI.

The best Time to cut Timber....................................... 72

CHAPTER XII.

Importance of a Supply of Wood................................... 75

CHAPTER XIII.

Preservation of Forests.. 78
Management of Forests... 79

CHAPTER XIV.

Establishing new Forests.. 80

CHAPTER XV.

Forest Trees... 87

CHAPTER XVI.

Evergreen Trees... 233

CHAPTER XVII.

Coniferæ or Cone-bearing Trees.................................. 237

CHAPTER XVIII.

Additional list of Coniferæ...................................... 278
Trees not Generally Known....................................... 282
Additions and Corrections....................................... 283

PREFACE.

The preface of a book is usually considered the proper place for an author to give his reasons for writing it. Following the usual custom in this matter, I may say that I am a son of a carpenter, who followed the business of building bridges, barns, houses, and similar structures, and my earliest recollections take me back to the time when I spent many an hour in the shop, twirling and unrolling the long, silky pine and white-wood shavings, and at these times I heard discussions almost daily in regard to wood, timber, trees, their quality, value, and variety. My father also owned a farm in the heavily wooded regions of Western New York, and he highly appreciated the value of certain kinds of trees growing thereon, for his practiced eye would measure the size of a hewn stick of timber that could be made from a giant oak, beech, or cther kind of tree as it stood in the forest, as well as make a very close guess as to the number of feet of boards or plank that could be produced from the great white-woods, hemlocks, or pines, of those regions. Brought up amid such surroundings, and early taught to use tools and work in wood myself, it was but natural that I should take an interest in Forestry, and endeavor to learn something of the value of trees and forests.

A few years later, or in the summer of 1846, I spent several weeks in the great pine forests of Eastern Michigan, commencing at Port Huron, at the foot of Lake Huron, thence travelling northward to the Straits of Mackinaw. This extensive region was at that time an almost unbroken wilderness, although there were a few saw-mills scattered here and there along the lake shore, or in the bays, that afforded a good harbor for the small vessels engaged in transporting lumber. The mills at Port Huron, Saginaw, Thunder Bay, and a few other places were kept running, but they made only a slight impression upon the surrounding forests, and it was often asserted at that day, that the pine forests of Michigan were simply inexhausti-

(5)

ble. But when a few weeks later I crossed Lake Michigan, and
travelled across Southern Wisconsin into Iowa, and in returning passed over the great prairies of Illinois, I began to realize
the fact, that while there were great forests not far distant,
there were also still more extensive regions of country that
would and must be supplied with lumber and timber of various
kinds.

A nine years residence in Wisconsin and Illinois, and several
journeys across the Great Plains, west of the Mississippi Valley,
with rambles in the Rocky Mountain regions, both in summer
and winter, have added something to my acquaintance with
our great forests, and strengthened my convictions as to their
importance and value to the country. During all these years I
have been engaged more or less in raising and planting forest
trees, sometimes as a business, but frequently as a pastime, or
for the purpose of experimenting with the different species,
both exotic and indigenous. Twenty-five years ago I commenced writing about forest trees, and from that time to the
present, I have never allowed a season to pass without urging
upon our people the importance of not only preserving the forests
we now possess, but also the necessity of planting new ones.

In 1864, at the urgent request of my former publishers, I
wrote a little hand-book called the " Forest Tree Culturist,"
which was to be issued in a pamphlet form and sold at a low
price, but after it was out of my control, these former publishers saw fit to add a cover of cloth, and offer it at the same price
as my larger works, a change that I have always regretted, as
it was not just to the author or purchaser.

Many a time during my life have I felt the need of some one
volume of moderate size, containing the names and descriptions, however brief the latter might be, of all the trees indigenous to the United States. Having waited in vain for the
appearance of such a book, I have attempted to write one myself, with the hope that it will be of service, not only to those
who may desire to raise forest trees for pleasure or profit, but
to others, who, like the author, may occasionally visit different
parts of the country, and need some such guide, that will
help them to call to mind the names, as well as assist in identi-
fying the different species of trees to be found in our forests.
I have written it for those who are not supposed to have given
the subject of forestry any special attention, and for this reason purposely avoided using any greater number of scientific

and unfamiliar terms than was actually necessary in describing
the various species and varieties.

Furthermore, at the suggestion of the publishers, I have been
as brief as possible, in order to make a book that can be sold at
a price within the means of all, and one that will not even deter
the summer tourist, who is about to spend a few days or weeks
in the country, from dropping a copy into his grip-sack before
leaving home. Could I have followed my own desires and
pleasures in this matter, the book would have been extended to
a thousand pages, and illustrated at a cost of many thousands
of dollars, but there are comparatively a small number of persons who take sufficient interest in forestry to purchase such a
work; consequently my own wishes have been made subservi-
ent to these circumstances, which neither author nor publishers
have power to control.

I desire to acknowledge my indebtedness to the excellent botanical works of Prof. Asa Gray, and those of the late Dr.
Chapman, of Florida, and Dr. Engelmann, of St. Louis, also to
those other botanists, which I have had occasion to consult, and
especially to the recently completed "Botany of California," by
Profs. Gray, Brewer, Watson, and their many able assistants.
To the works of Michaux, Nuttall, and other earlier writers I
have frequently referred in the following pages, also to the
"Book of Evergreens," by Josiah Hoopes, to which I have
called especial attention in the Chapter on Coniferæ. I have
endeavored to give proper credit to the first describer of the
species, but may in some instances have failed in this, owing to
the confusion existing in regard to this matter in some of our
botanical works.

ANDREW S. FULLER.

Ridgewood, Bergen Co., N. J., 1884.

PRACTICAL FORESTRY.

CHAPTER I.

INFLUENCE OF FORESTS ON CLIMATE.

The influence of forests on climate, is a subject that has attracted the attention of all civilized nations, and barbarous races, however low in the scale of intelligence, know enough of the effect of forests on climate to seek or avoid them as may be necessary to escape disease, or obtain shelter. We are not, however, to suppose that large forests are always a blessing to a country or to a people, or that their total absence is in all cases a dire calamity, for the jungles of India, or the almost impenetrable forests of the tropical regions of America, are no more desirable as places of residence, than the arid plains of this or any other country. What man should seek, is adaptation of the climate to his needs, and if he can increase or decrease the amount of moisture by changing the area covered by forest, he should lose no time in beginning to raise trees, or to destroy them, which ever is likely to conduce most to his welfare.

The cutting down of great forests, thus allowing the air and sun to reach the earth, and the wind to sweep over its surface, must necessarily hasten the disappearance of moisture therefrom, just as the opening of the windows of a room tends to a more rapid movement of the air within, and aids in dispelling smoke, steam, or odors which it may have previously contained. We ventilate a building by arranging for the ingress and egress of air in such a way that it shall be kept in motion, and we do the same thing on a more extended scale, when we raise or remove forests. Whenever an extensive region of country is denuded of its forests, the winds pass over it with greater velocity, impinging with greater force upon the soil, rapidly dispelling the moisture on it or arising therefrom. Keeping this in mind, we can readily understand why a country denuded of its forests may become so dry as not to admit of the production of grain, or any of the ordinary cultivated crops, while the annual amount of rainfall remains almost if not quite as great as it was when the forests were standing, and when the husbandman seldom failed to raise remunerative crops.

Any one who has resided for any length of time on the plains of Colorado, New Mexico, or in fact almost anywhere in the elevated regions of the West, can fully understand the effect of winds on surface moisture deposited by rains. I have known more water to fall in one hour in these regions, than in any four I ever experienced in the Eastern States, while owing to the compact nature of the soil it could not penetrate to any considerable depth, but passes over to the lower lands and streams, leaving the ground nearly as dry as before it came. These showers are almost invariably followed by high winds, which take up and dispel what little moisture may have remained attached to the leaves and stems of the low-growing weeds and grasses. The immense number of deep gullies to be seen all through what has been aptly termed the "arid belt," show plainly enough that very heavy showers do occasionally fall in these regions, but there is no large area of sponge-like leaf-mould, in either forest or field, to take up and retain the waters until utilized by plants—dispersed by slow evaporation, or absorbed by the soil beneath, where a portion of it at least would find its way to the little springs below, which in turn would feed the brooks and streams of lower levels. Instead of these natural obstructions, the way is clear for the most rapid departure of all the water that may fall in the form of rain. A few showers at a certain season, may produce a great amount of water, and still the section be so dry as to be almost uninhabitable the remainder of the year. One-half the quantity of water, if distributed through a longer period, might be all that was actually necessary to make the soil fertile and the climate delightful.

In many instances the destruction of large forests appears to have diminished the amount of rainfall, while in others no diminution has been observed. Col. Playfair, British Consul for Algiers, in a report to the home Government, instances some remarkable effects of extensive destruction of forests in that country. "During the first twelve years, since 1838, from which time meteorological observations have been carried on in Algiers, the rainfall averaged 32 inches annually. During the second twelve years it had decreased to 30.8 inches, and during the last fourteen years, it has been but 25.5 inches. The decrease became apparent after the principal clearings of wood in 1845, and in 1876 so exhausted had the soil become, that a famine seemed imminent in Western Algiers."

Similar instances in the decrease in the amount of rainfall following the destruction of forests, have been reported by several observers in various parts of the world, but principally by those residing in hot climates. Wherever forests of any considerable extent have been destroyed in Australia, Africa, India, Ceylon, or in the islands of the Indian and Atlantic oceans, lying within what may be termed the tropical belt, drouths seem to have almost invariably followed. These drouths, however, have not in all instances been traceable to a diminished amount of rain, but to rapid dispersion of moisture by winds, as well as evaporations from a soil exposed to the direct rays of a tropical sun. In fact, all written history

that gives us any information relating to the clearing of
the earth's surface of forests for the use, convenience, or
other purposes by man, show that it -diminishes atmos-
pheric moisture more or less.   In some instances this may
be beneficial, especially in regions where there is too
much rain and moisture for the convenience and prosperity of the inhabitants.   For this reason, it cannot be
said that the destruction of forests is always to be depre-
cated and looked upon as an evil, for it may be a blessing
in more ways than one.

The healthfulness of a country is frequently influenced
by the condition of the forests.   It has often been claimed
by those who are supposed to be acquainted with such
matters, that the draining of swamps, pools, and even
the under-draining of arable lands, tends to increase the
healthfulness of a country or neighborhood.   While this
may be true in some instances, the dispersion of what
may be termed surplus moisture, does not always produce
desirable results.

For a hundred years or more, it has been noticed that
the climate of the Island of Mauritius was changing from
one of great humidity to one of extreme dryness,
which, it is claimed is due to the destruction of the great
forests that originally covered the country.   So great
has been this change, that large tracts of land once
occupied by sugar plantations have been abandoned in
consequence of the severe drouths, which are of such
frequent occurrence that planters will no longer take the
risk of planting cane, while at the same time the mor-
tality among the inhabitants from fevers has been very
great, and these diseases appear to increase in severity as
the humidity of the climate decreases.   It is quite
probable, however, that the fevers are due to the drying
up of the surface moisture, and to the stagnant pools
formed during the rainy season.

When the country was covered with forests, and rains

were more frequent and abundant, the soil was shaded
and covered with leaves and other vegetable matter,
that prevented the rapid evaporation of moisture—a
portion of which must necessarily have found its way by
percolation into the ponds. The water in these ponds
was not only frequently renewed by showers, but pu-
rified by almost constant aëration, in consequence of the
visits of water fowls that frequented them in search of
food. Every ripple of the surface forced air beneath it,
and the movements of the birds—reptiles, amphibia, and
aquatic insects making their homes, or occasionally
visiting these ponds, assisted in aërating and purifying
the water. Ponds under such conditions never give off
fever germs, no matter in what country or climate they
may be located, but when they dry in consequence of
scarcity of rain, or the water is removed artificially,
there is always more or less danger of the emission and
dispersion of fever germs.

The draining of swamps, ponds—the changing of the
beds of streams—opening of new streets, even in our
older cities, as well as the breaking up of the virgin
soils of woodlands or prairies, are operations very likely
to be followed with outbreaks of chills and fever, among
the inhabitants of the neighborhood. The principles
that appear to govern the developement of fever germs,
are the same the world over, and if ponds, swamps, or
low lands are to be drained at all, it should be thoroughly
done, that there shall be no repetition of the danger
which usually follows the first disappearence of the
water therefrom.

We naturally look to the old world for information
in regard to the influence of forests on climate, because
there men have paid the most attention to the subject,
at least in modern times, and even if we go further back
and grope about among ancient cities buried in drifting
sands, or pass over desert wastes, where once forests

stood, while near by, fields of waving grain rejoiced the hearts of the husbandmen, the lesson is the same,—forests, fields, and firesides, are three inseparable links in the golden chain of man's prosperity as a tiller of the soil.

But the destruction of forests in the old world interests us, mainly as a warning, showing what may happen in this, if we continue doing as we have done during the past half century, in stripping the land of forests. I doubt if we have any proof that the destruction of forests thus far, in America, has had any perceptable influence upon the amount of rainfall, and there are not wanting instances where more rain has fallen in the open country than in the forests, but I believe that the fact is well established that in wooded countries, or where forests abound, it rains oftener and the atmosphere is, in consequence, more humid than where the opposite conditions exist. Marsh in his "Man and Nature," sums up this question of the effect of forests and rainfall as follows : "The effect of the forests then, is not entirely free from doubt, and we cannot positively affirm that the total annual quantity of rain is diminished or increased by the destruction of the woods, though both theoretical considerations and the balance of testimony strongly favor the opinion that more rain falls in wooded, than in open countries, one important conclusion, at least, upon the meteorological influence of forests is certain and undisputed ; the proposition, namely, that within their own limits, and near their own borders, they maintain a more uniform degree of humidity in the atmosphere, than is observed in the cleared grounds, scarcely less can it be questioned that they promote the frequency of showers, and if they do not augment the amount of precipitation, they equalize its distribution throughout the season."

There are, no doubt, great irregularities which must be

taken into consideration when making observations in regard to the influences of forests on climate. There may be long series of years in which drouths. will prevail, even in close proximity to very extensive forests, and these may be succeeded by seasons in which an unusual amount of rain will fall, but these extremes occur in all countries, and they do not prove that the average amount of moisture during a longer series of years has been increased or diminished by physical changes of the earth's surface, wrought by the direct agency of man. It must not be supposed that a few acres, more or less of forests, will produce any appreciable effect on the climate of the surrounding country, but they may, and in fact, usually do have a local influence in preserving humidity as well as affording shelter and protection against prevailing winds.

FORESTS AND STREAMS.

In all forests there is more or less vegetable matter, made up of leaves, twigs, old wood, mosses, and decaying herbaceous plants, all of which go to make up a sponge-like mass, covering the earth and filling the interstices between rocks, or perhaps such has been washed into depressions where the land is uneven or much broken up. But in whatever position it rests it absorbs and retains a large amount of water that falls in rains, or is produced from melting snow, until it slowly sinks into the soil below or is dispersed by evaporation. A part of that which passes into the soil is taken up by the trees, and exhaled by their leaves, thereby adding humidity to the surrounding atmosphere ; another part passes beyond the reach of the roots, and finding subterranean channels is carried onward until it again comes to the surface in springs, or sinks to some lower depth and entirely disappears.

That the vast deposits of vegetable matter in our great

forests are the reservoirs from which innumerable springs and brooks are supplied, is unquestionable, and not only are there hundreds of instances on record of springs and brooks drying up in consequence of the destruction of adjacent forests, but there are few persons who have reached middle life, that cannot call to mind more than one such, with which he has had personal cognizance.   If the little streams cease to flow through the greater part of the year, it must necessarily effect the larger ones. In all regions where there is considerable snow in winter, it remains much longer in the woods where it is shaded, than upon the bare hills and mountains, hence, the more continuous flow of brooks that have their source in elevated forest covered regions.   If the trees are removed from the hills, mountains, and elevated regions of a country, the great masses of vegetable mould which absorbs, retains, and checks the rapid descent of water from the higher to the lower levels disappear, and instead of water falling upon a sponge-like bed it strikes the bare earth or rocks from which it slides, rushing onward with constantly increasing velocity—forcing brooks and rivers to overflow their banks, often causing great destruction of life and property.   In the rapid movement of water from higher to lower levels, it removes all the lighter and more fertile parts of the soil, and this is repeated until the mountains and hillsides have lost the last remnant of a fertile soil, and become totally barren. Such lands can never be of any great value for cultivation, and for this reason, if no other, they ought to be reserved and kept covered with forests, as part of the public domain.

If forests tend to increase the rainfall of a country, as has been quite generally claimed, it might seem paradoxical to assume that they could in any manner have the least influence in preventing floods, for the more rain, the more water to escape and pass off in our streams,

and while it cannot be urged that the preservation of
forests, however extensive, will insure a country against
the recurrence of disastrous floods, they certainly do have
a modifying influence on the water that flows from the
higher to the lower levels, and finally reach the brooks
and larger streams.   Before any considerable amount
of water can pass from forest-covered regions, the great
deposits of vegetable matter covering the land must
necessarily become saturated and then only will there be
an overflow, besides the leaf-mould, sticks, brush, logs,
and similar materials, which are more or less abundant in
all forests, aid in retarding the flow, even after the
absorption has ceased—hence, we can readily understand
how a large volume of water may be held in check, and
prevented from a rapid descent to the streams below.
The leaves, twigs, and rough bark on the larger branches
and stems of the trees, also intercept the rain falling
upon them, and thus diminish the amount of water that
would otherwise reach the earth.

TREES FOR SHELTER.

Pioneers in heavily-wooded regions are usually anxious
to make a clearing, and as every tree felled not only increases the area which he is to cultivate, but extends his
view, the axe is often kept in use long after there is any
necessity for the purpose of obtaining land for cultivation.   In a few years the settler, who was at first so
anxious to open up the country, finds he has gone a little
too far in this direction for his own comfort and that of
his animals, for on taking down the screen he has not
only admitted the cold winds of winter, but those of
summer sweep over his fields, driving away needed moisture—whip the fruit from his trees before it is ripe, and
otherwise causes loss that might have been prevented.
It is then that he begins to feel the need of protection,

and to wish that his house and outbuildings were located by the side of some friendly forest or grove.

But if the inhabitants of once thickly wooded regions feel the need of shelter, how much more must those who settle in the prairie regions, where there are tens and even hundreds of miles, over which the wind sweeps at all seasons, without so much as a shrub to interfere or check it in its movements. It is in these treeless regions that forests are needed for giving shelter to man and beast, and also to protect the fields and orchards of the husbandman. Forests are the natural remedy for the imperfections of the climate of the prairie region, and while they may not do away with all the objections that might be urged against such regions, they certainly go far towards ameliorating present conditions. The remedy is a simple one, and not beyond the means of the poorest. Trees are cheap, and can be as readily grown as the most common vegetable of the garden, when one has learned how do it.

FORESTS AND INSECTS.

Forests were, without doubt, the original home of some of our noxious insects, but they were also the home of their natural enemies, among which we may safely place in the front rank the insectivorous birds. But when the forests are destroyed, the birds seek a home elsewhere, or are destroyed or frightened away by hunters, and while the insects may in a measure be disturbed, they still find food in our orchards, gardens, and among ornamental trees of various kinds. Give the birds shelter and treat them kindly, and they would in many instances aid us in keeping down our insect pests. It is true there is a difficulty in distinguishing friends from foes among the birds, and even when we are able to do this, it is scarcely possible to drive away our enemies without at the same time frightening our friends.

CHAPTER II.

THE CHARACTERISTICS OF TREES.

The trees of the world are separated by botanists into two grand divisions, known as exogens or outside growers and endogens, or inside growers. These two divisions are also called dicotyledonous and monocotyledonous, the first having two cotyledons or seed-leaves, as seen in the sprouting acorn or young maple (fig. 1) the two lower leaves being the cotyledons), and similar tree seeds, while the others have but one cotyledon or seed-leaf, as seen in the cocoanut, date, and other species of palms. As we have but two or three arborescent species of the palm, and a yucca or two that reaches a hight of even small trees, and these are of no especial value, I shall have no further occasion to refer to monocotyledonous plants in the ensuing pages. All of the ligneous or trees with firm wood, belong to the exogens, but in some in-

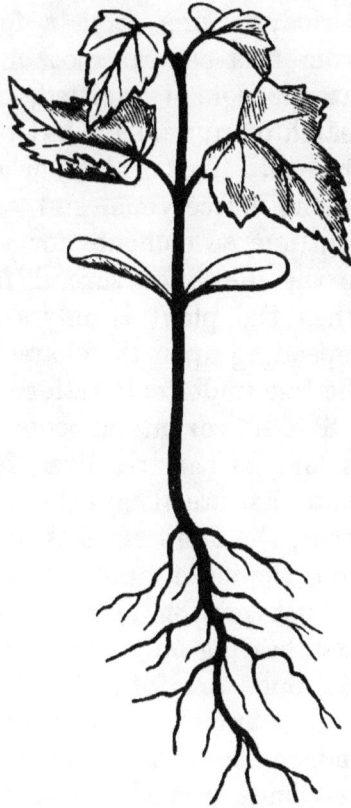

Fig. 1.—SEEDLING MAPLE.

stances, such as the pines, the embryo is provided with more than two cotyledons, and there are from three to ten seed-leaves instead of two, but there are never less than two. As the seedlings grow up into

trees, their stems and branches increase in diameter by
the annual formation of a new layer or ring of wood, deposited on the outside of that of the preceding year—
hence the name of the outside growers.

The root at first is but a single descending axis, growing downward and absorbing nourishment from the surrounding soil, for the support of the ascending axis or
stem. This condition or form of root exists for only a
period, varying from a few hours to a few days; for,
from this central root or radicle of seedlings, side or
lateral roots are emitted, not only as it would appear in
search of nutriment, but to more firmly fix the plant in
the soil. This central, or as it is more commonly termed
among nurserymen and arboriculturists, tap-root, may
continue to elongate for years, and penetrate the earth
to the depth of several feet, or it may cease to grow
when the plant is only a few weeks or months old, all
depending upon the character of the soil, or the habit of
the tree under cultivation.

The side or lateral roots, however, continue to elongate
as long as the tree lives, for it is the newer or younger
roots that are always the most active in absorbing nutriment, the more rapid their development and multiplication, the more rapid is the growth of the whole tree.

That portion above ground is at first but a simple
stem bearing only leaves, but as it increases in hight and
age, buds are formed on the central axis and from these
springs branches, and this multiplication of branches and
buds continues throughout the entire life of the tree. The
first buds formed on the stem may or may not produce
branches, depending on circumstances, but as a rule only
a few of the uppermost do so, and the others are overgrown and smothered.

As a whole, a tree may be said to consist of roots,
stem, branches, buds, leaves, flowers, and fruit, with
bark surrounding all the ligneous parts. This bark is at

first very thin, but subject to great alterations with age, owing to the distention through the increasing diameter of the stem, as well as the formation of new layers of liber or inner bark. This annual addition from within pushes outward the older bark, often causing it to crack open, forming deep fissures in the outer surface, or to fall off in scales.

The stem of the tree is composed of wood in different conditions. The term alburnum is applied to the new or sap-wood, through which the crude sap absorbed by the roots passes upward to the leaves where it is assimilated. In returning, it is distributed over the entire surface of the tree, forming new layers of wood and bark. Some authors have applied this name to the half-formed vegetable matter, lying between the bark and the wood during the growing season, but that is now called by vegetable physiologists the "cambium layer," to distinguish alburnous or fully formed young wood. This alburnum or sap-wood, sooner or later, is mostly changed into heart-wood, assuming in most kinds of trees a dark color as seen in the red cedar, black walnut, beech, and oak, and although it is in fact dead wood, decay is prevented through its protection from the air, by the surrounding layers of alburnum.

Some kinds of trees have very thin layers of alburnum, especially those of slow growth, while the stems of others appear to be all sap-wood, as seen in the white pine, tulip tree, and white-heart hickory, but this is more in appearance than reality, the difference in color between the old and new wood being but slight. The change from soft-wood to heart-wood is not sudden, but proceeds slowly, the cell walls gradually becoming thicker and more rigid with age, and the difference in color is due mainly to chemical changes. The alburnum or outside layers decay when exposed to the air far more rapidly than the heart-wood; consequently it is less valuable for posts, rails, shingles, or other articles used in outside work, but it is sometimes used for inside finishing of buildings and furniture, giving greater variety. The young or outside layers of wood are the toughest and most flexible, as the filling up or thickening of the cell wall of the heart-wood makes the timber more firm and rigid, and doubtless more durable, but at the same time its elasticity and toughness is diminished.

That the inside or heart-wood is dead, and only serves to strengthen the tree mechanically, is shown in the fact that it may be removed entirely by decay, and still the tree grow on vigorously for centuries. This leads me to the subject of

THE MOVEMENT OF SAP IN TREES.

All plants obtain their nourishment in a liquid or gaseous form, by imbibition through the cells of the younger roots or their fibrils. The fluids and gases thus absorbed, probably mingling with other previously as- similated matter, is carrried upward from cell to cell through the alburnum or sap-wood until it reaches the buds, leaves, and smaller twigs, where it is exposed to the air and light, and converted into organizable matter. In this condition a part goes to aid in the prolongation of the branches, enlargement of the leaves, and formation of the buds, flowers, and fruit, and other portions are gradually spread over the entire surface of the wood, ex- tending downward to the extremities of the roots. We often speak of the downward flow of sap, and even of its circulation, but its movement in trees in no way corre- sponds with the circulation of blood in animals, neither does it follow any well-defined channels, for it will, when obstructed, move laterally as well as lengthwise or with the grain of the wood.

The old idea that the sap of trees descended into the roots in the fall and remaining there through the winter, is an error with no foundation whatever. As the wood and leaves ripen in the autumn, the roots almost cease to imbibe crude sap, and for a while the entire structure appears to part with moisture, and doubtless does so through the exhalations from the ripening leaves, buds, and smaller twigs, but as warm weather again approaches, and the temperature of the soil increases, the roots again commence to absorb crude sap and force it upward, where it meets soluble organized matter changing its color, taste, and chemical properties.

If this was not the case, we could not account for the saccharine properties of the sap of the maple, or for the presence of various mucilaginous and resinous constituents of the sap of trees in early spring, because we find no trace of such substance in the liquids or crude sap as absorbed by them from the soil. If the growth of a tree continues all the season without check, there will be one well defined ring of new wood deposited over the entire outer surface; but in some instances drouths check growth in mid-summer, and these being followed by heavy rains and warm weather, a second growth often takes place, producing a second deposit of new wood. In what may be termed cool climates, it seldom occurs that a second deposit is of sufficient thickness to be distinguished from the first, and as a rule the age of a tree may be determined by the annular rings, provided, of course, they are sufficiently distinct to be counted.

THE BUDS OF TREES.

For all practical purposes the buds of trees may be divided into four classes, the terminal, axillary, accessory, and adventitious. What are usually termed fruit buds by horticulturists, may be placed in the second division, because they have not generally a fixed character, but are analagous to a leaf bud, and while under

favorable conditions they develop into flowers, under
others they merely produce leaves or their axis is extended into a branch.   The terminal buds, which crown the
apex of a stem or the ends of branches, consist of undeveloped leaves, which only require an elongation of the
stem to allow for their full development.

Axillary buds are those on the axil of each leaf, on
the small twigs or on the yearling stems of seedlings,
and from these the branches or lateral shoots are produced.

Accessory buds are merely a multiplication of the axillary bud, two, three, or even more in a cluster, but it is
seldom that more than one of the number develops,
the others remaining dormant.

Adventitious buds are those which may be developed
from almost any part of the stem, and are in no way dependent upon any natural location of leaves, joints, or
internodes.   In some plants, like the willows, poplars,
hickories, and chestnut, they may appear from wounds
on the stems of large trees or from exposed roots, in
fact, the cells of some kinds of trees appear to possess
an inherent property, which enables them to become
buds or roots, according to the conditions under which
they are placed.   Such plants are usually considered as
very tenacious of life, or as having great vitality, while
those of an opposite nature are far more difficult to propagate, and require more care to cultivate.   But this
may be due in part, at least, to our ignorance in regard to
what they require for their full development, under artificial conditions.

CHAPTER III.

RAISING TREES FROM SEED.

Naturally, seeds drop from trees directly to the ground, or are scattered by the winds for some distance. To provide for their wider distribution many kinds of seed have their membraneous appendages or winged margins. These are termed Key-fruited. The maple, elms, birch, ash, and tulip trees are familiar examples among the deciduous trees, and the common arbor-vitæ, pine, and spruce among the evergreens or coniferæ. The seed of oaks, hickory, and nut-bearing trees generally, are not scattered any considerable distance from the parent stock, except through the agency of mice, squirrels, and other small animals, who carry them away for food, but occasionally leave them in a position conducive to future growth. It might be supposed that Nature would make no mistakes in placing seeds in the best possible position for germination, and were she at all chary in regard to waste, we might find it so, but being prodigal in all of her productions, the preservation of one seed in a thousand or even in a million is sufficient for her purpose. Or we may look upon this seeming extravagance as purposely intended to supply with food the hordes of animals that are known to live on seeds—the perpetuation of the species being dependent on what is left, after the animal creation have been well supplied. But we can readily see that a large portion of all the seeds that fall do not find congenial places for growth, even if they be not interfered with by animals or man, for some drop in stony places, others upon the dead leaves, where they dry up and wither. Seeds of the large fleshy-fruited trees, as the apple, pear, plum, or oranges, lemons, and

2

the like, are enclosed in substances, the decay of which usually causes the death of their germs.

With these examples before us, we can readily say that while Nature "doeth all things well " for her own use, man can and has improved upon her methods for supplying his own needs ; therefore, in raising trees from seed, we follow Nature's guidance only so far as her ways serve our purpose. We do not scatter acorns and hickory nuts over dry leaves in a forest, and expect them to grow more readily and better than when planted in a good artificially prepared soil and covered a proper depth. But with other kinds we find shade, which young seedlings receive as they come up in the forests essential, hence, we are obliged to provide it, when the same kinds are raised under wholly artificial conditions. These variations are mentioned here, because there are certain would-be teachers of arboricultural science, who are continually holding up Nature or natural methods of propagating trees as the only true ones, and deprecating any departure therefrom.

The seed of all trees and shrubs grow readily under what may be called artificial conditions, and we have only to take cognizance of their distinctive characteristic and provide for the same, in order to be successful in growing them. Seeds that are small and enclosed in a thin shell or husk, should not be kept for any considerable time in a dry atmosphere before planting, or buried deeply in the soil. The large and coarser kinds will withstand more exposure and ill-usage—but even these respond promptly and generally to good treatment. As the space at my disposal will not admit of giving specific directions on the management of all the different species and varieties of trees in cultivation, I can only give briefly general rules and methods for raising trees and shrubs from seed.

In the region where they are grown, or in similar latitudes and climates, all kinds of seed may be sown so soon as ripe. But when taken from one locality or country to another, variations in the time of sowing should be made to correspond with the change of climate. Although the proper, or natural time for planting seeds would seem to be immediately after their ripening, it is frequently impracticable to plant at such times, and is seldom done by those who make raising trees a specialty, for if placed in the soil in autumn, mice, moles, and other animals are very likely to attack and make sad havoc with them before the growing season returns, and in hard tenacious soils the earth will often become so firmly packed over the seed during the winter, that the young sprouts frequently fail to break through in the spring. In light friable soils, and where there are no vermin to destroy them (which is rarely the case), most kinds of tree seeds may be sown in the fall.

There are, however, two species of our native maples, Scarlet and Silver-leaved, and also the different species of elms, the seeds of which mature in the spring or early part of the summer, and as they usually germinate soon after falling (at least those of the maples do), they should be sown as soon as ripe. But these are exceptions to the general rule.

The seeds of a large proportion of deciduous trees may be preserved over winter by mixing them with clean, sharp, moist sand, and burying in the ground, covering only just enough to protect them from vermin and the changes of weather. A dry knoll or other well drained situation should be selected. Acorns, chestnuts, and and hickory nuts and seeds of the later ripening maples, locusts, three-thorned acacia, yellow wood, and hundreds of other similar kinds will keep perfectly in this way, and be found in excellent condition in spring, when they may be sown with, or without the sand in which they have been stored.

While all of the larger nuts, as well as maples, tulips, elms, magnolias, and several other kinds, will not germinate after becoming once thoroughly dry, there are other kinds as the locust, yellow wood, acacias, and nearly all of the coniferæ that may be kept in a cool, dry atmosphere a year or more, and some will retain their vitality for several years. There is also another class of trees, the seeds of which can scarcely be forced to germinate until they have been in the ground for two years, among them the common hawthorn (*Cratægus*), and the red cedar (*Juniperus*), and closely allied species are familiar examples.

PREPARING A SEED-BED.

For most of the deciduous trees the open field is a good situation for a seed-bed, no shade being required for the young seedlings, except in rare instances. The preparation of the soil should be most thorough, not only should it be plowed deep, but cross-plowed and pulverized with a harrow, until in fine tilth and free from all lumps and stone. If the land is not rich it should be made so,˙ by liberal applications of very old and well decomposed barn-yard manure, or some other good fertilizer, but no fresh stable manure or other kind that will make the soil too open and loose, should be used. When all is ready, the seed should be sown in drills far enough apart to admit of cultivation, with plow or cultivator. There are two methods of sowing, the single drill and in double or narrow beds. The first is more convenient for thorough and clean cultivation, but the latter is sometimes preferred, where the space to be devoted to the purpose is limited, or where it is desirable to raise a very large number of plants on a given area. Small seeds may be sown with a seed-drill, when convenient, or the single drill may be opened with a plow or marker made for the purpose, or even opened with a hoe drawn along by the side of a line for a guide. The depth of the trench must be varied according to the size and kind of seed to be sown, For maple, ash, locust, and similar kinds one-half inch of soil is sufficient covering, but the larger nuts should be covered a little deeper. Judgment should be used in all cases, and the depth of covering be varied not only with the size of the seed but with the nature of the soil. If this is light and sandy, or contains so much vegetable matter that it does not become compact, and the surface hard after heavy rains, the seeds may be covered deeper than in one of an opposite character.

On sowing in what are termed double trenches or narrow beds, a trench a foot wide and of proper depth is

Fig. 2.—THE TREE DIGGER.

opened, the soil being thrown upon one or both sides. The seeds are then scattered on the bottom of the trench, and the soil drawn back over them.

The wide drills should be three or four feet apart, or at sufficient distance to admit of pruning and cultivation between them, and to give room for workmen to pass when hoeing and weeding the plants. Frequent stirring of the soil between the rows with plow and cultivator during the summer materially increases the growth of the plants, as well as facilitates the emission of side or lateral roots. At the end of the first season, or certainly not later than the second, the plants should be dug up. This may be done very rapidly with spades, or faster and better with a tree digger represented in fig. 2. This very handy implement passes under the plants, cuts off the tap-root if long, and at the same time leaving them standing upright in the row, from whence they can be readily pulled up by men following the digger, or left to be taken up when wanted. Having used one of these implements for many years, I can speak from experience of its value, especially for lifting seedlings that have very long and coarse tap-roots, like the black walnut, hickories, and similar kinds.

After the plants have been lifted, the long tap-root should be shortened if it has not been cut off with the digger. Some of the nut trees, like those mentioned above, will throw down a central or tap-root to the depth of two to three feet the first season, while the stem above may not be more than a foot high. Fig. 3 represents an average specimen of a one-year-old seedling black walnut. The tap-root of such a plant should be cut off at *a*, and the larger lateral roots going below this point either spread out or shortened.

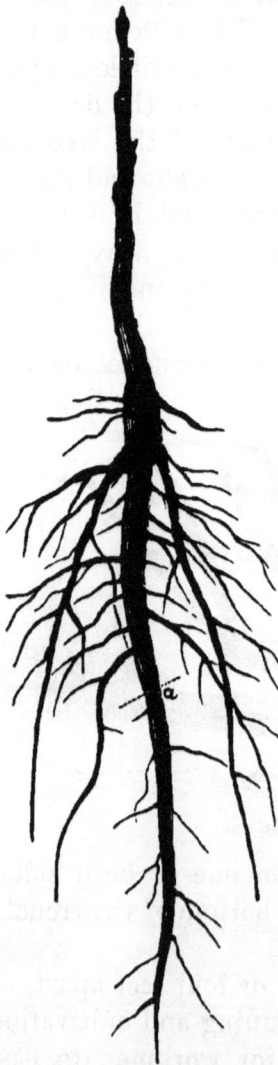

Fig. 3.—SEEDLING BLACK WALNUT.

The main object in shortening the tap-root is to force
out side or lateral roots the following season, but it also
renders transplanting less troublesome, as it would be
very inconvenient to dig trenches or holes three feet
deep in which to set seedlings not more than one or two
years old. These tap-roots are doubtless of value to
trees growing thickly in the natural unbroken soil of a
forest, and where there is little room for side or lateral
roots to grow, without coming in contact with those of
neighboring trees, and where it is necessary for roots to
go deep to find moisture, as when growing on high and
dry soils, but it is seldom that trees growing sparsely or
in low moist soils retain their tap-roots many years, if
they have them at all. Therefore they can only be considered necessary appendages under certain conditions,
none of which often exist in cultivated trees.

I am well aware, that there are arboriculturists in this
country who will not agree with me in this, for some
often claim that the central or tap-root is a very essential
part of a tree, and for this reason they advocate planting seeds where the tree is to grow in order that it should
be preserved intact. But with all due deference to the
opinion of these gentlemen, my long experience with
trees has shown me that tap-roots are but short-lived at
best, except in rare instances, and only with trees growing on dry, hard soils, where all the roots go down deeply in order to reach moisture. I have taken up thousands of trees from moist soils and of all ages, from one
to twenty or more years old, and I never found one with
a tap-root of any considerable size, and generally there
was none at all on trees after they had reached the age
of a half dozen years or more. I have also seen hundreds of acres of our largest forest trees turned out by
the roots by tornadoes, and by stump-pullers in clearing
the land for canals and railroads, but not one in a hundred had anything like a tap-root.

CHAPTER IV.

TRANSPLANTING SEEDLINGS.

The seedlings of forest trees raised under artificial conditions should always be transplanted while young, and generally at the close of the first season, or when one year old. In some instances where only a feeble growth has been made, or the kinds are of a dwarfish habit, the plants may remain in the seed-bed two or more years before removal, but as a rule, the first transplanting should be made earlier. In cold climates, and in soils where the frost is likely to lift the plants, or otherwise injure them, they should be taken up so soon as the first frosts have killed the leaves, and heeled-in where they can be protected from cold.

In warm climates the transplanting may be direct from the seed-bed to the nursery rows, or to the grounds where the trees are to grow, but it is seldom advisable or safe to set out small one-year-old seedlings in a forest where cultivation is not practicable, the better method being to set in nursery rows and give good cultivation for a few years before planting them where they are to remain permanently. When in nursery rows they are in a convenient position for training into any desirable form, and their roots will be materially increased in number by the frequent stirring of the soil in which they are growing.

Nursery-grown trees, and those that have been frequently transplanted while young, are re-transplanted with less labor and more certainty of living, than those that have not passed through these preparatory stages. I know of no tree that is at all difficult to transplant, if it has had proper culture while young. The hickories, tulips, and magnolias are generally considered the most sensitive of all our native deciduous trees, but by beginning when they are young, and subjecting them to root-pruning as directed, and repeating the transplanting every three or four years, large masses of fine fibrous roots will be formed near the main stem, that will insure their successful removal, even when the trees are fifteen or twenty years old.

Every time a tree is lifted from the earth in which it is growing, the ends of the fibers or larger roots are broken off, or at least disturbed to an extent that prevents their further elongation when again placed in contact with the soil, but fibers push from the sides, and thus the number of roots is increased within the radius occupied by the longest, or those extending farthest from the main stem. It is by this transplanting and shortening of the leading roots that the arboriculturists are enabled to produce a mass of close compact fibrous roots that are easily preserved when removals are necessary or advisable, and the life of the plant is not endangered by the operation. For the reasons given, nursery-grown trees, or those raised under artificial conditions, are much better for planting than those that come up naturally in forests, but the latter can be subjected to the same preparatory operations and made valuable, if it is commenced while they are young, or of moderate size. The treatment, however, to which forest seedlings should be subjected must vary somewhat according to the kind of tree, as well as the character of the roots, the latter often differing greatly in different soils. For instance, seedling trees found growing in low, moist soils, seldom have very long central or tap-roots, while on all dry uplands, the contrary is the case, as I observed on a preceding page.

If the seedlings when taken from their natural habitats have an abundance of small, fibrous roots, they may be treated in the same manner as those from nurseries, the tops being pruced to give the plants the proper form.

It is always best, however, to prune seedlings from the
forests a little more severely than those from a nursery, as
the former will feel the change more on account of having
been moved from a half shady position to one fully exposed
to the sun.   The amount of stem and branches to be left
on seedlings obtained from the forest, must always be in
proportion to the quantity and quality of the roots; if
the latter are few in number and weak, then but a small
part of the stem should be allowed to remain.

As an extreme of what may be considered severe
pruning to insure success with seedling trees taken from
the forest, I will cite my own experience in handling
several thousand tulip trees (*Liriodendron Tulipifera*).
Desiring to procure a number of these trees, more for
experiment than anything else, I sent my workmen to
the woods and adjoining fields that were partly overgrown
with brush, to get the required number.   Finding the
seedlings had but few side roots, and but one or two
long tap-roots reaching down into the subsoil, the spades
were thrown aside and the trees pulled up with what few
roots might adhere to them.   They were of varying sizes,
from two to eight feet high, with stems from the size of a
pencil to an inch or more in diameter.   The roots were
almost entirely destitute of fibers, and resembled carrots
more than the roots usually found on trees.   All the
larger trees were cut down to one foot, and the tap-root
shortened to about the same length, or a little less.
These stumps were planted in a light, sandy soil, in
nursery rows, and given the usual cultivation, with a
loss of less than five per cent.   As the sprouts started,
all except the strongest one nearest the top were rubbed
off.   Some made a growth of two feet or more the first
season, and the next all would average three feet in
hight.   The short stump above the point where the
new sprout started from the main stem was cut off
smooth during the summer; the wound soon healed

over, and was entirely obliterated in the next year or two. The third season the trees were transplanted, and the roots found to be very numerous and in excellent condition. A few hundred of these trees I retained, and set them out in nursery rows four feet apart each way, where they remained three years more, at which time they were ten feet high, with large spreading, handsome heads, and a mass of roots that would have been pronounced perfect by the most exacting arboriculturist. A number of these trees are now growing in sight from my library window, and I am quite certain that they are not only larger, but more healthy and beautiful than any of their companions of the same age left in the forest from which they were taken.

During the removal of trees from forests or nursery, it is quite important that the roots should not be exposed to drying winds or to the light, more than is necessary, and they need not be so exposed over five minutes at the most, when being dug up or set out. The roots should be covered so soon as taken from the ground, and kept moist until set out again, whether they are transported a long or short distance. A good way to protect the roots of trees, is to coat them with thin mud, or puddle them as it is termed. A few moments' time spent in making a mud hole, into which the roots are dipped as dug, or soon after, will often be the means of saving them and making a success of what might otherwise prove a failure.

The preceding remarks relative to pruning both roots and branches, are applicable to deciduous trees only. Evergreens require different treatment, and will be considered in another place.

CHAPTER V.

BUDDING AND GRAFTING.

BUDDING.

The propagation of woody plants by the process known as budding, consists in taking from one tree or shrub, a bud and transferring it to another. The plant upon which the bud is placed is called the stock. The limits of this operation are not very well defined, but for all practical purposes I may say that it is limited to the members of the same genus, or closely allied plants; that is, oaks may be budded on oaks, chestnut on chestnut, and generally the nearer related the species, the more successful the operation. But like all other rules pertaining to the propagation of plants, there are exceptions, and occasionally we may find that the wood of two species belonging to the same genus, cannot be made to unite and form what is termed a union. There is always a preference in stocks belonging to the same genus, and the propagator seeks the best for his purpose. I may say, however, that as a rule, the weak and feeble growing should always be placed upon the strong growing, if rapid growth and long life is the object of propagation.

Budding is usually performed in summer, soon after the buds or a portion of them are fully developed on the young wood of the present season's growth. The stock into which the buds are to be inserted must be in a similar condition, although the stem or branch at the point of junction may be more than one year old, but in no case must the bark be so thick and rigid, that it cannot be readily separated from the wood beneath, because the bud is to be inserted under the bark of the stock, and unless this can be done the operation will fail. We have

to depend upon the assimilated or true sap to form a
union between the bud and the stock, the same as we do
on layers and cuttings to produce roots, for all the operations are analogous, only in budding, the alburnous
matter forms a union with the same material in the
stock, while in the layer and cutting, it is emitted in the
form of roots.

The proper time for budding trees must, of course,
vary with the latitude, season, and kind of trees to be propagated, as some come forward earlier than others, but, as
a rule, it can be performed as early
in the season as good plump buds
can be found at the axils of the
leaves in shoots of the present
season's growth. The upper and
immature ones can, of course, be
discarded, if it is necessary to
commence budding before all are
in fit condition for use. In fig. 4,
a, we have a bud which is to be
transferred to a stock; a knife is
inserted about one inch below it
and passed upward, and brought
out about a half inch above, cutting out a piece of bark with a
thick slice of wood of a form
shown by the circular line in
the figure. We now make a cut across the stock, cutting
just through the bark, and another longitudinally downward, as shown in fig. 5, then insert the lower end
of the bark containing the bud, under the bark of the
stock at the point where the incisons meet, and press it
down to its place. If the bark of the stock is firm, and
does not part easily to admit the bud, the edges must be
lifted so as to allow the bud to pass under it freely. If
the piece of bark containing the bud does not pass com-

Fig. 4.
THE BUD.

Fig. 5.
THE STOCK.

pletely under, then cut it off at the upper end even with the cross-cut in the stock, so that it will fit in smoothly. In fig. 6 a bud is shown, taken out after the upper end has been cut off, as directed, and on this is also shown a portion of a leaf-stalk, usually left attached for convenience in handling the bud, as well as to protect it from injury. After the bud is inserted, it is secured in place by a ligature, which may be of bass bark, a strip of thin cloth, woollen yarn, or any similar material that will hold the bud and bark in place, until a union is formed. The point of the bud and leaf-stalk attached should, of course, be left exposed. The stock into which a bud is inserted should not, as a rule, be over an inch in diameter or less than a half inch, although much larger and smaller are often used. After the bud has firmly united with the stock—which will usually be in two or three weeks—the ligature should be loosened or removed entirely. The bud is not expected to push into growth until the following season, at which time the stock above the bud should be cut away and the bud allowed to grow undisturbed. If sprouts appear on the stock they must be removed, in order that all the strength may go into the bud.

Fig. 6.
THE BUD REMOVED.

The horizontal incision in the stock is sometimes made below or at the bottom of the perpendicular one, and the bud thrust under the bark, but upward, or the reverse of the more usual method, this permits the downward flow of the sap to reach the bud in a more direct course than when the cross-cut is made above it. It is not a convenient method, but is sometimes desirable when the flow of sap is rather sluggish, as it often is late in the season.

When a bud is taken from the shoot in the usual way, there is a small slice of wood remaining under the eye, which, in budding some kinds of plants. it may be de-

sirable to remove, although it is an almost universal
practice in this country to allow this wood to remain,
and doubtless in a majority of cases, it is best to do so;
but there are instances where a more permanent union
will be secured if it is re-
moved. With kinds of trees
like the magnolias, horse-
chestnuts, and common sweet
chestnut, that have a rather
thick bark on the young
shoots, better success will be
attained by the removal of the
wood from the bud. When
this is to be done, the shoots
used must be in a condition to
allow the bark to peel readily
from the wood, without tear-
ing or breaking the fibers.
Hold the branch in the left
hand with the smaller end towards you; insert the knife-
blade about one inch below
the bud, cutting a little deeper
than you would if the wood
were to be left in, pass the
knife under and above the bud,
some three-quarters of an inch,
but not out to the surface, but
withdraw the blade, and cut
across through the bark only
about a half an inch above
the bud, then with finger and
thumb lift up the bark, at

Fig. 7.—TAKING OFF THE BUD.

the same time press it gently forward, and you will remove the bark and bud (fig. 7) *a*, without injuring it,
leaving the piece of wood *b*, adhering to the branch.

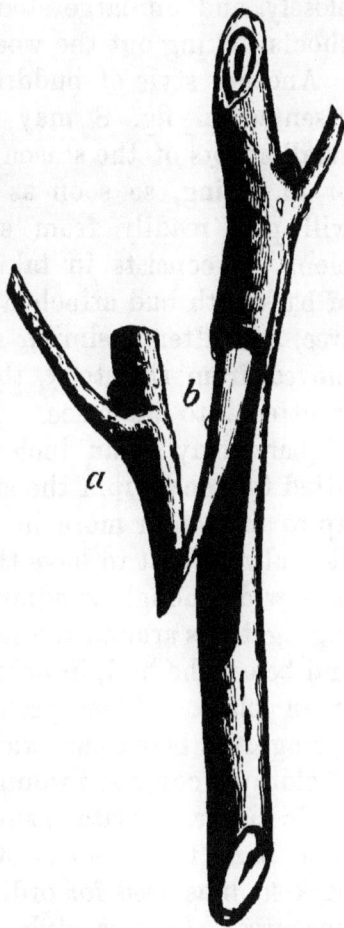

This is a much better and more scientific method of removing the wood than to pick it out with the point of a
knife, or to remove with a goose-quill as sometimes recommended. This concave piece of bark, with the bud
attached, will fit the convex surface of the stock very
closely, and on large stocks, and with buds from large
shoots, taking out the wood is often advisable.

Another style of budding called the annular, and represented in fig. 8 may be practised in summer on
small shoots of the season's growth
or in spring, so soon as the bark
will peel readily from stock and
cion. It consists in taking a ring
of bark with bud attached from one
tree, and after a similar ring is removed from the stock, the former
is fitted into its place. This ring
of bark may be an inch wide and
fitted to stocks from the size shown
up to an inch or more in diameter.
It is always best to have the ring of
bark wide enough to admit of placing ligatures around the stock above
and below the bud, in order to hold
it in place. When performed in
spring, it is best to use waxed strips
of cloth, to cover the wound and exclude the air, but late in summer and

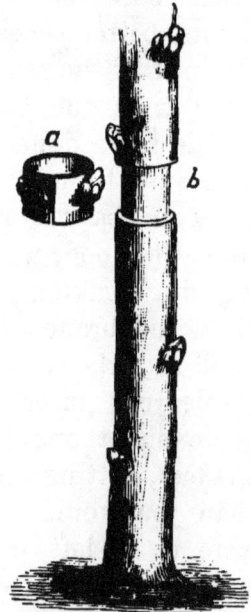

Fig. 8.
ANNULAR BUDDING.

with bark from shoots of the present season, strips of
bark such as used for ordinary budding, will answer for
ligatures. In this style of budding, the branch from
which the ring of bark is taken, should be nearly the
same size as that of the stock to which it is affixed.

In performing these operations an implement called a
budding-knife is required, and they are made of various
sizes and patterns, and are usually to be obtained at al-

most any seed store. The imported budding knives have usually either a thin,
blunt-pointed ivory, or bone handle, or a
piece of bone inserted into a horn handle,
this being used to lift the bark of the
stock, to facilitate the inserting of the
bud under it. Many gardeners and
nurserymen still use these old forms of
budding knives, but they are clumsy
affairs, and not adpated for rapid work.
Any small pocket knife with the blade
rounded, and made thin and smooth,
will answer fully as well for the purpose
as the most costly imported ivory-
handled knife. I have never seen a
knife that I liked better than the one
shown full size in figure 9, which I first
saw in use at the old Linnæan Nurseries,
at Flushing, N. Y., some thirty years
ago. Unfortunately, however, these
knives are not in the trade, and when
wanted have to be made to order. But
by purchasing cheap knives at the hard-
ware stores, and throwing away the
blades, and have new ones put in, such
knives do not cost any more, or in fact
quite as much, as the rugular trade budding-knife. The rounded end is used
for lifting the bark on the stock, and
far more convenient than a knife with
an ivory handle, which must be reversed
in the hand every time a bud is inserted,
and this is a waste of valuable time, besides the ivory or bone handles are far
more likely to become rough, and scratch
the tender cambium layers than a piece of polished steel.

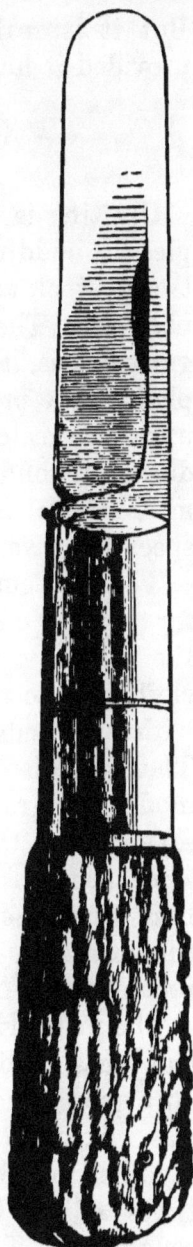

Fig. 9.
BUDDING KNIFE.

But it is really immaterial what form of knife is used, provided it has a keen edge, and is dexterously handled.

GRAFTING DECIDUOUS TREES.

Grafting is governed by the same physiological princi. ples as budding, and there must exist an affinity between stock and cion, if not, a permanent union is impossible. The principal difference between budding and grafting, is, that in the latter a larger section of the plant to be propagated is used, and it can be performed upon a greater variety of plants while they are in a dormant condition. The art of grafting is one of the most ancient methods known of multiplying individual species and varieties of plants.

The implements used for grafting are : a small saw for cutting off the heads of large stocks or branches of the trees, a good strong knife with thick back to make clefts in the stock ; a small knife to prepare the cions with ; a wedge, grafting chisel, and a small mallet. There are also many other implements used for different modes of grafting, but they are really not essential, except when the operator desires to cut a cion or cleft of some peculiar form. In addition to the implements, bass strings, such as are used in budding, for tying in the grafts, or grafting-wax, to cover the wounds, and protect them from air and water are necessary.

The primitive composition used for covering wounds and cuts made in grafting, was clay and cow manure. Any good kind of clay was taken, and two parts of this was mixed with one part of cow manure, all well beaten together in order to make it as tough as possible. Sometimes a little finely chopped grass was added to give it toughness. This composition was in common use more than two thousand years ago, and is still used by gardeners in grafting certain kinds of plants that have a soft, sponge-like bark that might be injured by compositions containing oil or grease.

The composition of grafting wax is almost as variable as the ideas of the men who use it, and there are scores of recipes for making it. One of the oldest, and, we think, one of the best for out-door use, is composed of four parts of common rosin, two parts of beeswax, and about one and a half parts of tallow ; all melted together over a moderate fire, and well stirred before the mixture cools. If it is to be used in very cool weather, add a little more tallow, or if in warm a little less. In Europe, Burgundy pitch is more generally used in making grafting wax than in this country. Some of the French nurserymen recommend the following : Melt together two pounds twelve ounces of rosin, and one pound and eleven ounces of Burgundy pitch. At the same time melt nine ounces of tallow, pour the latter into the former while both are hot, and stir the mixture thoroughly. Then add eighteen ounces of red ochre, dropping it in gradually and stirring the mixture at the same time. After the composition has cooled sufficiently, work it well with the hands. For out-door work in cool weather this wax is rather hard, but if carried in a vessel where it can be occasionally warmed it is readily applied, and is quite durable.

All the above kinds of wax may be spread upon cloth or tough paper with a brush when warm, and after it has cooled the paper or cloth may be cut up into narrow strips of any convenient size. In what is called splice or whip grafting, these strips of waxed cloth are very convenient for wrapping about the parts united. A French mastic known as " Lefort's Liquid Grafting Wax " is made by melting one pound of common rosin over a gentle fire and adding one ounce of beef tallow, the latter to be well stirred in. Take it from the fire, let it cool down a little, and then mix in eight ounces of alcohol.

The alcohol will cool down the mixture so rapidly that it may be necessary to put it on the fire again. The utmost care must be exercised to prevent the alcohol taking fire. This mastic is highly recommended by the nurserymen of France, and it has been used to some extent in this country for several years. It is imported in tin boxes, and usually kept on sale at seed stores.

I might give many other recipes for making grafting wax and mastics, but believe that the above are the best, and that the one made simply of rosin, beeswax, and tallow is as good as any ever invented. Some nursery-men of late years have substituted linseed oil for tallow, and while it may answer for some kinds of trees, I am inclined to think it is injurious to those with very thin bark. I have known several instances where losses have occurred that were attributed to the use of oil in making the wax in grafting.

In all the different modes of grafting, great care should be observed in having the external surface of the wood of the stock and cion, to be exactly even, no matter whether the external surface of the bark is even or not. This allows the new cells, which form between the bark and wood, of both stock and cion to unite and form a channel, through which the sap can readily pass. The sap ascends through the wood of the stock into that of the cion, causing the leaves to expand, which, in their turn assimilate it, preparatory to its return to stock and roots below.

The time for grafting most kinds of woody plants in the open air is in the spring, just before or at the time the sap begins to liquify, varying the time to suit differ-ent species, for experience has demonstrated, that there are some which may be operated upon much earlier than others. The shoots or young twigs to be used for cions, may be taken from the parent stock in autumn, and pre-served in earth, charcoal, sawdust, moss, or some similar

material, where they will be cool—not frozen—and just
sufficiently moist to prevent shrivelling. Cions of ripe
wood may also be cut at the time they are used, but their
vitality is often weakened by the severity of the weather,
and their delicate tissues injured to such an extent, that
they will not form what is called in grafting " granula-
tion," (although it is precisely the same as the callus on
cuttings), which fills up any small interstices that may
exist between the stock and cion, allowing a communica-
tion between. Wood of one season's growth is preferable for cions to older (except in rare instances), and it
should always be firm and fully matured, and selected
from the most healthy and vigorous branches. As there
are many hundreds of different modes of grafting, I
shall only mention a few of the most simple, because the
difference between the larger part is so slight, as to be
scarcely worthy of a different name.

Cleft Grafting.—This method is principally used upon
large stocks or on the branches of old trees. The stock
is first cut off at the
point where it is de-
sirable to insert the
cion; it is then split
with a large knife or
chisel, being careful
to divide the bark,
and at the same time
leave the edges
smooth, as shown in
fig. 10, when the

Fig. 10.
CLEFT GRAFTING.

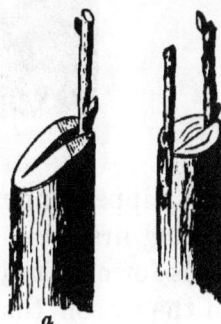

Fig. 11.
CLEFT GRAFTING.

knife is withdrawn, the cleft may be kept open by inserting a wedge made of iron or hard-wood. The cion (fig.
10, *a*), should be two or three inches long, bearing at least
two good buds. The lower end is cut wedged-shaped as
shown, in order that it shall fit the stock. In stocks of
an inch or more in diameter, two cions, one on each side

as shown in fig. 11 may be inserted, and if both grow, one can be cut away. In stocks of less size, one cion will be sufficient, and the top of the stock will be cut off with an upward slope as shown in fig. 11, *a*. After the cions are inserted, the entire exposed surface of the wood should be covered with grafting wax or waxed cloth.

Crown Grafting.—This is but a mere modification of the cleft craft, but instead of splitting the stock to receive the cion, the latter is sloped off thinly on one side

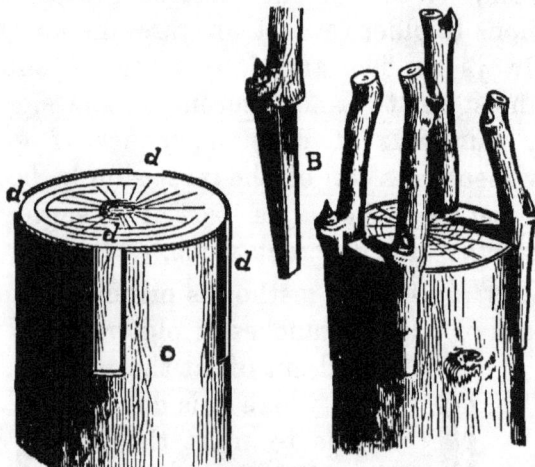

Fig. 12.—CROWN GRAFTING.

and slipped under the bark, as is done in budding, a slit having first been made in the bark of the proper length. This form of grafting is usually performed a little later in the season than the last, in order that the bark may be separated from the wood of the stock. The cions used are cut earlier in the season, and kept dormant in some cool place until wanted for use. Another form of crown grafting is shown in fig. 12, the cion is cut about half-way through as shown, and the wood removed, leaving a square shoulder at top and opposite to a good bud. From the stock *d, d, d, d*, the bark is removed to admit the cion, and one to four cions as shown, are

fitted to a stock, and then all held in place by ligatures of waxed cloth, and the top of the stock also covered
with wax. This mode of grafting is practised on very
large stocks such as are not suitable for cleft grafting.

Side or Triangular Grafting.—This is a modification
of cleft or crown crafting, and instead of splitting the

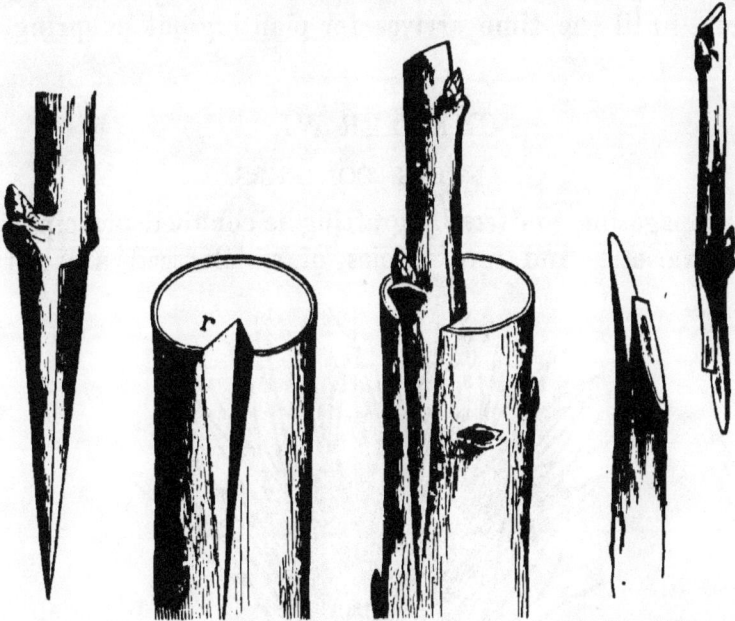

Fig. 13.—TRIANGULAR GRAFT.

Fig. 14.—SPLICE OR
TONGUE GRAFTING.

stock, a triangular incision is made in the side of the
stock, as shown in fig. 13, *r*, and the cion cut in the same
form and fitted into the cleft as shown.

Splice or Tongue Grafting.—When the stock and cion
are nearly of the same size, splice grafting is the most
convenient and certain method known. The stock is cut
off with an upward slope, and a small cleft or split is
made in it, about midway on the slope, forming a tongue.
The cion is cut in the same way, but with a downward
slope, with a corresponding tongue, and the two are then

neatly fitted together, the tongue on one entering that of
the other, as shown in fig. 14. Ligatures of waxed cloth
or strong paper must then be applied to hold the cion in
place. This is a convenient and rapid mode of grafting
small stocks or roots in the house or open ground, and is
largely employed in grafting the apple during the winter
months, the grafted plants being packed away in sand or
earth, until the time arrives for planting out in spring.

CHAPTER VI.

GRAFTING CONIFERS.

Propagating conifers by grafting is confined principally to varieties and rare species, of which seed are not

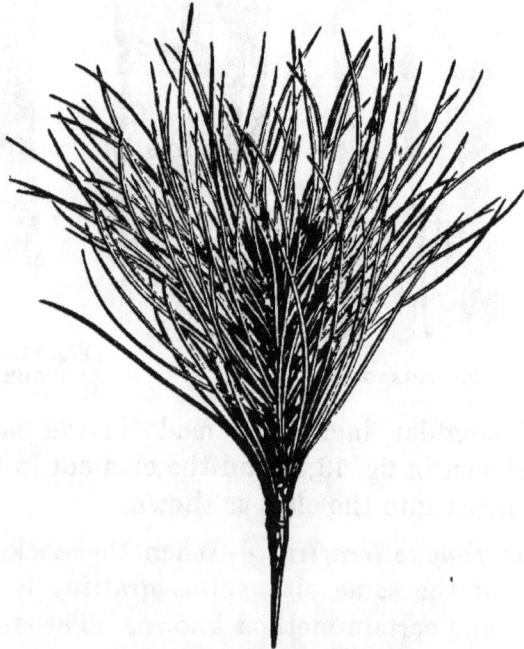

Fig. 15.—CION OF PINE.

readily obtainable, or that are not readily multiplied by
cuttings or layers. In this country, grafting of evergreens is usually confined to plants raised under glass,

or those placed in frames for the purpose, where the grafter can control the temperature and supply moisture as required, until a perfect union between stock and cion has taken place. Grafting conifers in the open air may sometimes be done quite successfully, especially upon some of the species of arbor - vitæs, yews, larch, and taxodiums, but as a rule it is best to use small stocks, and have them potted in the fall, and then keep them in a cool green-house or frames, and then graft as they commence to grow late in winter or early spring, the cions being either in a dormant condition, or the buds but slightly advanced.

The cions should always be of the

Fig. 16.—GRAFTED PINE.

previous season's growth, and a portion of the leaves left attached. The stocks also must not be denuded of their foliage, and a convenient method of grafting is what may be termed a side graft, the cion being inserted into a cleft, made in the side of the stock, and held in place with the usual ligature of bass bark as in budding. The plants are kept in a somewhat confined atmosphere, and frequently syringed overhead until the cion has united, then the stock above it is cut away.

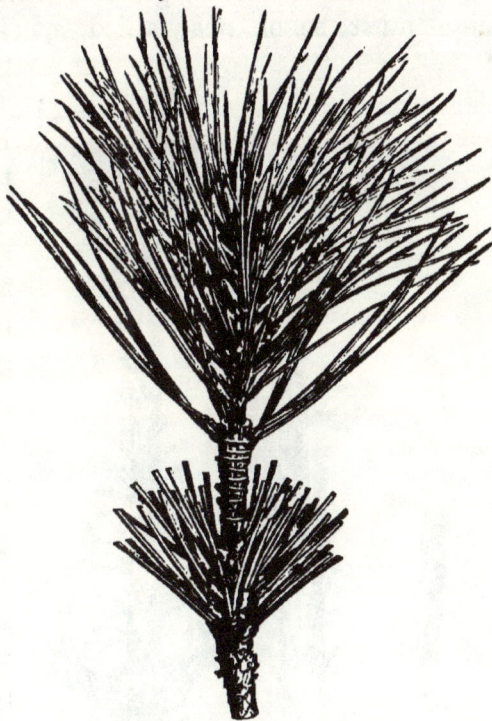

3

Evergreen trees that do not belong to the coniferæ, can also be grafted quite successfully in the same way, in fact, usually are given the same or similar treatment. The cions may be one to three inches long or even longer in some instances, varying according to the species propagated. Of course, in grafting conifers the stock and cion must be of near allied species, the pines may be worked on pines, spruce on spruce, etc. It is always advisable to select a strong and vigorous growing species as a stock for a weaker one. What is called terminal grafting is sometimes practised quite successfully, and in fig. 15 is shown a cion of pine prepared for inserting in a stock, and in fig. 16, the same is shown in place, and fastened by a narrow ligature. The leaves at the point where the cleft is made in the stock

Fig. 17.—TERMINAL GRAFT.

are wholly removed, while a few below are shortened to allow of applying the ligature as well as inserting the graft.

Another mode of terminal grafting as sometimes employed on the balsam fir, is shown in fig. 17, the cleft being made in the end of a shoot, dividing the terminal buds, and the cion inserted between as shown.

The deciduous conifers. like the larch, taxodiums, salisburia, etc., may be multiplied quite rapidly by grafting on pieces of their own roots, or those of closely allied species, and in the same manner as other trees are root-grafted, and during the winter months.

CHAPTER VII.

CONIFERÆ FROM CUTTINGS.

There are many species and varieties of the coniferæ that are readily propagated by cuttings. It is practised extensively with species of which seed cannot be obtained, also with varieties upon whose seed little dependence could be placed, of producing from them plants like the parent tree. Varieties of evergreens are no more likely to come true from seed than varieties of the pear or apple; consequently more direct methods of propagation must be adopted. The arbor-vitæs, junipers, yews, tor-reya, cephalotoxus, podocarpus, cryptomerias, and species belonging to several other genera, are quite readily prop-agated by either cuttings of ripe or green wood. Some will grow quite readily without artificial heat, especially in the Middle and Southern States, but success is more certain if all are placed where the temperature can at all times be under the perfect control of the propagator. In warm climates, a mere frame covered with glazed sash, or thin cloth, may answer, but in cold ones it is better to have some means of giving the cuttings a little extra heat during the winter months. An ordinary cool green-house is perhaps, all things considered, the best kind of structure in which to propagate evergreens from cuttings, especially in cold climates. Cuttings are made of the ends of the smaller branches, and mainly of the ripe one-year-old wood, but with some kinds a little of the two-year-old may be taken at the base of the cutting.

The cuttings should be of good size, that is from three to four inches long, and the leaves on the lower half cut away, and the lower end of the cuttings made smooth, a sharp knife always being used for this purpose. Fig. 18 shows a cutting of arbor-vitæ prepared for planting, and fig. 19 one of the Lawson cypress (*Cupressus Lawsoniana*). Sand is usually preferred to soil in which to rear cuttings of evergreens, and it may be put in boxes of convenient size for handling, or in larger frames or on benches fitted up for the purpose, but boxes will usually be found most convenient, as it enables the operator to change his cuttings from one place to another, should it be found necessary to secure a proper temperature. The boxes used may be four or five inches deep, and eighteen inches or two feet square, and when filled with moist, sharp sand, they are ready for

Fig. 18.—ARBOR-VITÆ CUTTING.

use. To make a channel in which to set the cuttings, use the edge of a pane of window glass, sinking it to the proper depth in the sand, and straight across one side of the box. Set the cuttings in this close together, until it is filled, press down the sand firmly against them, and then make another crease in the sand, about two inches distant from the first, and proceed in this manner until the box is full. Apply water to further settle the sand about the cuttings, then place the boxes in the shade until roots are produced, which in some cases will take six months, while in others they will appear in a less number of weeks. Our object in all such cases is to give the cuttings a chance to throw out roots before the top is forced into growth, as will usually follow placing the cuttings in full light and in a warm atmosphere.

In all cases where ripe cuttings are employed for propagating evergreens, time must be given for the cuttings to become well furnished with a callus on the lower end, before they are forced into growth, else they are certain to fail. Sometimes the cuttings are kept through winter in a moderately warm room, and in spring placed in a hot-bed, where they will receive bottom heat to assist in the production of roots, and forcing a growth of the tops. The propagator can always learn how his cuttings are progressing, by taking out a few occasionally and examining the condition of the callus at their base. If after they have been planted two months or more no callus is to be seen, he must give a little more heat, or if

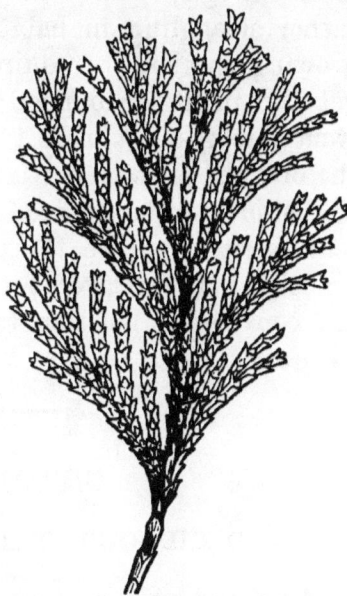

Fig. 19.—CUTTING OF CYPRESS.

they are in a cold frame in the open ground, add a little more covering. Small evergreens are sometimes potted and kept in a green-house during winter, and when they have made a new growth, this is taken off for cuttings, which grow very quickly when placed in a confined atmosphere, and a high temperature. But such methods of propagation are seldom practised, except by nurserymen who have all the necessary facilities for the rapid multiplication of the different kinds of tender as well as hardy plants.

LAYERS.

All the different species of evergreens that can be propagated from cuttings may also be layered in the same manner as directed for deciduous trees, but it is seldom practised to any considerable extent, except with dwarf and trailing species. Plants produced by layer are usually rather straggling in habit, and if of naturally upright growing species, it requires more pruning and care to get the plants into good form than with those raised from cuttings. A mere twisting, coiling, or notching of the branch, so as to partially separate the wood and bark at the joint to be covered by earth, is usually all that is required to increase the production of roots. Several branches may be layered from one plant, or all that are in a convenient position to be bent down and covered with earth.

CHAPTER VIII.

DECIDUOUS TREES FROM CUTTINGS.

There are quite a number of different species of trees that are usually propagated from cuttings of the ripe wood, instead of from seed, as it requires less skill to multiply them in this way than any other. Among these I may mention the willows, poplar, buttonwood, a few of the maples, some of the alders, etc. Those which grow freely, like the willows and poplars, require no special preparation, and the cuttings may be taken off in spring or fall in warm climates, and immediately planted out, and the cuttings may be of almost any size, from a few inches long to several feet, and be made of one-year-old wood,

or that which is older. But with some other kinds, like
the Negundo maple, and the buttonwood, the cuttings
should be made up in the fall, and from wood of the cur-
rent season's growth, cut into sections of from ten to
fifteen inches in length, and then heeled-in in some warm,
moist place, either in a cellar or in the open ground where
they will not freeze, and at the same time be kept sufficiently cool to prevent growth. In spring these cuttings may
be taken out and planted in trenches made with a plow or
spade, and deep enough to admit of covering the cuttings
nearly their entire length, and the soil should be pressed
firmly about them, after which they should receive the
same care as seedlings, in a similar soil and climate.

In making cuttings it is best to cut just below a bud,
and square across the wood, for the base of the cutting,
but the upper end may be sloping, although it will make
little difference, except with those kinds which have a
large pith, or those that do not produce roots very freely.
The object in making up the cuttings some months before
planting, is to give time for a callus to form on the exposed wood, a process that always precedes the emission
of roots from cuttings. It is really aiming at the production of roots in advance of the pushing of the buds
into growth, and while we may not always accomplish
this, we can at least secure a callus, which is a step gained
in the right direction. All such cuttings produce roots
more freely in a moist soil than in a dry one, and in dry
climates it is a good plan to cover the entire surface of
the cutting bed with some kind of mulch, in order to keep
the ground moist and cool during the summer months.
The cuttings, when rooted, should be treated the same as
trees raised from seed.

PROPAGATION BY LAYERING.

Layers are really nothing more than a form of cutting,
the only difference being that they are allowed to adhere

or remain attached to the parent stock—drawing suste-
nance therefrom until roots are emitted, after which they
are detached and become individual plants. In making
layers of trees or shrubs, we bend down a branch, and
cover that portion with earth upon which we wish to
produce roots. Fig. 20 shows a layered branch buried
in the soil. An incision is usually made on the under
side of the branch before it is laid down, and the knife
inserted just below a bud if there is one convenient, passing into the wood, and then an inch or more lengthwise,
the branch forming what is termed a tongue, as at a.
A hooked peg may be employed to hold the layer in place,

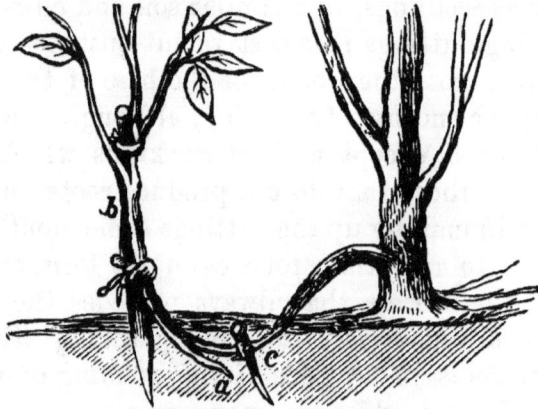

Fig. 20.—LAYER.

c, or a stone laid on it, as it is quite important the
branch should be held firm in place. If the branch is
large, the end may be tied up to a stake, as shown at b.
It is not often that forest trees, except some ornamental
varieties, are propagated in this way, but it is well enough
to know how to do it, when necessary to increase the
stock of some choice or rare specimen. Layers may be
made at almost any season, but they will root sooner if
made when the trees are growing rapidly, than at any
other time.

Some kinds of trees will produce roots when layered

without cutting of the branch, and exposing the alburnum—in fact, all will, in time, but the surest way is to cut the branch as described. With some kinds, roots will be emitted so slowly that the layer must remain at least two seasons before it will be safe to sever it from the parent stock. Evergreens may be layered in the same way as deciduous trees, but the operation should always be performed during the period of active growth, else the wound made on the layer is likely to be covered with rosin, which may prevent the emission of roots.

Sometimes a part of a tree or a small branch will vary from the original; when this occurs on a large tree and where the branch cannot be made to reach the ground, we are compelled to elevate the soil, or some similar material to the part we wish to propagate, unless it is some species which can be readily propagated from cuttings, buds, or grafting. If we desire to obtain a layer, we have only to place a pot or box of soil near the branch, so that it can be covered with earth, the same as if near the ground. After the branch is layered, the soil surrounding it must be kept moist until

Fig. 21.—A LAYER IN POT.

roots are produced. Fig. 21 shows a branch layered in a pot from which a piece has been taken from one side to admit the branch; this crevasse is closed with a piece of board or shingle placed on the inside of the pot, after which the pot is filled with soil. If the pot is surrounded with cloth or moss, it will in a measure prevent drying, and less frequent waterings will be required.

CHAPTER IX.

SEEDLINGS OF CONIFERÆ.

A large proportion of the cone-bearing trees are ever-greens, but there are a few, as the larch, and taxodiums, that are deciduous, casting their leaves in the autumn when fully ripe, or touched by frost. In propagating from seed, all require essentially the same treatment, which is, however, quite different from the ordinary deciduous class already referred to. While the seeds of coniferæ are really no more delicate, or their germination more uncertain than other kinds of tree seeds, still, the seedlings require more care from the time they appear above ground, until they are transplanted to the field or nursery rows. Young seedlings of coniferæ, that spring up in the forests, where there is deeper shade than that which surrounds those of deciduous trees, are quite sensitive to light, temperature, conditions of soil, and atmosphere, as regards moisture. As a rule, all seeds of conifers should be sown where the young plants can be protected from the constant direct rays of the sun for the first few weeks of their existence, and partial shade is desirable throughout the entire first season.

Sometimes stakes are driven by the sides of the beds, on which poles are placed to form a support for a cover-ing of thin cloth, or of evergreen boughs where they can be obtained, but in windy locations such temporary structures are liable to be blown down, destroying the plants underneath them. Besides the winds sweep over the surface of the seed-bed dispelling the surface moistures, which must be made good by liberal and judicious waterings, or the seedlings soon perish. Large numbers of seedlings are raised, it is true, under such arrangements in favorable situations, but a better protection is furnished by board frames and lath shades as shown in figure 22. The frames may be made of inch boards or plank, and should be one foot high, four feet wide, and of any desirable length. To make the lath shades, take ordinary ceiling lath four feet long, lay them parallel, and two inches or a little less apart, and fasten them in their places by nailing across their ends two strips of boards three inches wide and three feet long. This size of shade is more convenient for handling than larger ones, and being so light are not liable to be broken. These shades laid on the frames will admit air and moisture, and while admitting the direct rays of the sun to every part of the bed during the day, they will not

Fig. 22.—LATH COVERED FRAME.

remain long enough upon any place to cause injury. The constant change of the sun and shade thus secured, is just what is required by delicate seedling coniferæ, and some few other kinds of trees, that will be mentioned further on.

Seeds sown in frames as described are under the control of the cultivator. They can be watered when it is required, more or less shade given if desirable, besides being in a position to be protected in winter, should their hardiness be doubted. Occasionally it may be found necessary to increase the temperature of the seed-bed, or to protect the seed from long continued cold rains. This can be readily done by substituting for the lath, glazed hot-bed sash, that are usually, and should always be found as the ordinary adjuncts of every good garden.

The soil in which the seeds of coniferæ are sown should be of a light, porous nature, and if not naturally of this kind, it should be made so by liberal additions of leaf-mould and sand. If, after watering, the surface becomes hard, and a firm crust is formed, it is not light enough, and more sand or mould should be added. Pure sand will answer well for a sed-bed, if a little liquid manure be added occasionally after the seedlings commence to show their second set of leaves.

SOWING THE SEEDS.

Small evergreen seeds, like those of the arbor-vitæs, may be scattered broadcast on the surface, and then be covered by sifting soil over them, but those of the larger size should be sown in drills from four to six inches apart, and not so close in the drill that the growing plants are liable to be crowded. When sown in drills, the soil between them can be stirred if necessary, and the weeds more readily removed than when the seed is sown broadcast.

The time for sowing will depend much on the climate and the kind of seed. Such as can be safely kept through the winter, should be sown in the spring, although some of the more hardy species may be sown in autumn, but there is often danger of their destruction during winter by mice and other vermin. The usual course is to keep the seed over winter in their cones, as gathered from the tree, or if shelled out, in paper bags, stored in a dry, cool room. Should they seem too dry in spring, their germination may be materially hastened by soaking a few hours in tepid water. After the water is drained off, the seed may be mixed with dry sand or gypsum, which will take up the surplus moisture, and facilitate the ready separation of the seed when sowing. Very small seed should be covered an eighth of an inch, the coarser ones, like those of the nut-pine, a half inch, but not more.

Moisture and heat are requisites of germination; consequently, if the soil is dry at the time of sowing, or dry weather follows, water must be applied to the bed from time to time as required, but extremes in giving too much or too little must be carefully avoided. This is a nice point, and can be determined only by the experience, skill, and good judgment of the cultivator.

The seeds of most of our coniferæ germinate freely and quickly, but the young seedlings, when in what is termed the first or seed-leaf stage, fig. 23, are extremely sensitive to any considerable change of temperature or hydrometical condition of the air or soil, and, in fact, this is considered the critical period in the life of the young plant. If the weather is warm and wet, the soft succulent stems soon rot, or damp off, as it is termed. Dusting the surface of the bed and plants with fine dry sand, or what is still better, burnt and pulverized clay, will usually prevent further loss, unless the weather should continue unfavorable for many days. On the other hand, water must not be withheld in dry time for fear of causing the plants

Fig. 23.—SEEDLING PINE.

to rot, for wilting caused by want of moisture is almost as fatal to the plants as too much. After the second, or true leaves appear, the critical period may be considered past, and ordinary care only will be required during the remainder of the season. But it is well to apply water as needed throughout the summer. If the seedlings are raised in a cold climate some slight protection may be given during winter. It is not necessary or advisable to try to keep out the frost, but merely to spread something

over the frames to prevent the sudden and alternate
freezing and thawing of the ground. If snow comes
early, and drifts in and around the plants, this of itself
will be sufficient protection so long as it remains. If the
plants make a good growth the first season, or are likely
to be crowded during the second, they should be transplanted, but if there is room for them to grow they may
remain in the seed-bed two years and be transplanted in
the spring of the third season. But as soon as they are
large enough to handle conveniently, whether it be at
the end of the first or second season's growth, they should
be carefully lifted and set out in nursery rows as recommended for deciduous trees.

In most cases planting out may be done with a dibble,
as the roots of young seedling conifers are usually quite
small, but well furnished with fibers. When transplanting, great care should be given to the protection of the
roots from the sun, as well as to keep them moist. The
time to transplant is as early in spring as the ground will
permit of proper preparation, and even then a close exam-
ination of the roots will usually show that growth has
commenced, although the buds give no indication of the
movement of the sap. When first set out in nursery rows
the seedlings should be set not more than six inches apart
in the rows, but so soon as the branches of one plant touch
those of another, every alternate plant should be removed, or all taken up and replanted. The latter method
is the one usually adopted by nurserymen who desire to
make low stocky trees, with an abundance of fibrous
roots, because every time the tree is removed, the lateral
growth of the larger roots is checked, and new ones grow
from their sides. But the value of the trees and the purpose for which they are raised should be considered, and
their treatment in the nursery be in accordance with the
results desired. If they are to be grown for timber, hight
will be more essential than breadth in the young trees,

and the growth of their lower branches be discouraged, and the leading shoots preserved instead; but if the plants are intended for ornamental purposes, or to be set out as wind-breaks, screens, or in hedge-rows, low, stocky trees will be preferable, and the growth of the lower branches should be encouraged. An occasional cutting-in of the terminal shoots of the young trees, and transplanting every two or three years—allowing plenty of room for each to spread itself in all directions—will greatly facilitate the production of fine specimens for ornamentation, and that will fully meet the requirements of the planter.

The time at which to give the proper shape or the form desired is when transplanting, and with the exception perhaps of the pines, it may be done without regard to the position of the buds on the stems. The young trees may also be pruned at other times, but this will be done mainly by pinching off the young succulent shoots in summer, shortening those that push out to an unusual length, or at points where, if left unchecked, would give the tree an unsymmetrical form.

In removing evergreens from the nursery to forest plantations, a little more care is required than in handling deciduous trees, because the foliage of the former is always present, through which evaporation of the juices of the tree takes place to a limited extent, even during what is termed the dormant period, and the foliage suffers if the roots are long exposed to the light and a dry atmosphere. The holes made for the reception of the roots of transplanted trees should always be of sufficient size to admit of placing them in a natural extended position, and if considerable larger it will be better than to have them in the least cramped or crowded, and as a rule the roots should not be buried any deeper than they were before the tree was transplanted. Of course an inch or two of extra depth may be allowed for the usual settling of the recently disturbed soil.

EVERGREENS FROM THE FORESTS.

Vast numbers of evergreens are annually transplanted
from the natural seed beds that abound in many parts of
the country where and when the various species of conifers
are growing wild.   The source of supply is simply inex-
haustible, limited only by the number of desirable species
to be obtained.   Until recently, gathering natural seedlings of conifers was confined to some of the most common
species of the Eastern States, such as hemlocks, arbor-
vitæs, spruces, balsam firs, and pines, but within the
past few years, the facilities for obtaining rarer kinds has
been greatly extended, and the coniferæ of the Rocky
Mountain region and the Pacific Coast can now be obtained very cheaply, and collectors possessing the requi-
site botanical knowledge are employed to collect seedlings
of one, two, or more years old, and these are sent to different points, or wherever there is a demand for them, by
mail or otherwise.   When properly handled, these forest
seedlings will usually grow and make good specimens.
To insure success, the seedlings should be pulled when
the ground is wet, that as many of their roots as possible
may be retained, and these in no case should be exposed
to the sun or wind long enough to cause shrivelling.   If
packed in soft moss (*Sphagnum*) from low ground, they
can be safely transported in cool weather to any distance
not requiring more than six or eight weeks in their
transit.   When such plants arrive at their destination,
they should be unpacked and placed in a position where
new growth of roots can be secured without exciting or
forcing a new growth of the buds or branches.   This is
readily done, because roots will grow at a much lower temperature than buds, consequently it is only necessary to
heel-in the seedlings, or pot them separately if of rare
kinds and worth the trouble, and place them in a half
shady place, watering freely overhead until the roots

commence growth, then transplant or admit light and heat sufficient to insure a healthy growth. My own practice—and it has been eminently successful—has been, in the case of rare species, to procure them in the fall or early winter, and to place each plant in a flower pot of the required size, and then plunge them under the central stage of my green-house, or in frames where they would not freeze.

Seedling evergreens gathered in Oregon in November and received and potted the last of December, showed plenty of new roots by the first of March, while at the same time very few gave any signs of growth in their tops, beyond a slight swelling of the terminal buds. Out of two thousand obtained one autumn, and treated as above, the loss did not exceed five per cent. In handling seedlings of a foot or more in hight, the same idea as expressed above should be kept in view, and it is also well to prune away or cut back their leading branches before planting out.

When evergreen seedlings from forests near by are taken up in the spring, they should be set near together in rows or beds, and then well shaded until root growth has commenced, after which the covering may be removed entirely or in part. A convenient way to shade such plants, is to spread a thin layer of hay over the entire tops of the plants, and as showers beat down the hay from time to time during the spring and early summer, the foliage will be exposed more and more, and finally the hay will reach the ground and form a mulch for the plants during the rest of the season. Plants set near together, with rows eighteen inches or two feet apart, may be protected in this manner very effectually and satisfactorily.

Shading may be done with cloth, boards, or boughs of larger evergreens, or in any other way most convenient; but shade made in some way is usually necessary to insure success with forest seedlings. Those obtained from

open fields and along the borders of forests where they
have been somewhat exposed to light and winds, are
preferable to those grown in deep forest shade.

SEASON FOR TRANSPLANTING.

Volumes have been written advancing theories relative
to the proper season for transplanting evergreen trees and
shrubs, and while it may be possible to remove them safely
under favorable conditions at almost any time, it must be
apparent to every student of vegetable physiology, that
the proper or best time is when the plant is in its most
dormant condition.   But in cold climates this period is
so extended, that if the plants are transplanted at the
beginning of it, they become weakened and often entirely
deprived of their natural juices, by the evaporation
through their leaves and twigs before a fresh supply can
be obtained by the action of their roots upon the soil in
which they are planted.   The mere contact of roots with
moist soil is not sufficient to enable them to absorb
nutriment to any considerable extent, but contact through
growth is necessary to bring their absorbing functions into
action.   It is for this reason that transplanting evergreens
in cold climates should be done in spring.   If transplanted
early, or so soon as the ground is warm and dry enough
to work readily, there will usually be time for new roots
to form, through which sap will be imbibed to support
new top growth, as in the case of the potted seedlings
referred to on a preceding page.   The principles in both
cases are identical.   The large plant as well as the small
one needs time to become settled and fixed in its new
position before the growing season commences; consequently early planting is always preferable, and if new
growth of branches has commenced when the roots are
disturbed, it is very likely to be checked, even if the tree
does not die.

If for any cause it becomes necessary to transplant

evergreens after active growth has commenced, the rapid evaporation of moisture from the new shoots may in a measure be prevented by giving the tree shade. An old cloth, or the branches of other trees set around each one, will serve the purpose, and often be the means of saving the transplanted trees. Watering overhead is also very beneficial, for the same reason, and operates to check undue loss of moisture through the foliage and smaller branches. Evergreens that have been so frequently moved in the nursery that their roots form a solid ball, can, of course, be removed safely at almost any season, but such trees must be considered as exceptions. In all cool climates there are frequent showers during the spring months, and with the increasing heat of the sun, and an atmosphere laden with moisture, all nature awakes from its long slumber in a condition to recuperate, after having been disturbed or checked in its progress; hence it is the safest and best season in which to transplant trees of all kinds.

CHAPTER X.

PRUNING FOREST TREES.

All kinds of forest trees may be, and nearly all should be pruned at the time of transplanting. As it is almost if not quite impossible to take up a tree without destroying a portion of the roots, or at least disturbing them, it is well to reduce the number or length of the branches to fully compensate for any loss sustained by the roots. It is also better to prune away more wood than is actually necessary for the safety of the trees, than to fall short of removing enough ; for a few buds and leaves, fully supplied with nutriment, are worth far more to the tree, than a large number kept feeble for the want of it. I am

well aware that there are men who object to pruning
transplanted trees, because they imagine in their ignorance of the general principles of vegetable physiology,
that the larger the number of buds and leaves, the greater
the capacity of the tree to assimilate sap, which would
in a measure be true, provided the roots were in a condition to supply the crude article in unlimited quantities,
but as they are not at such time, it is absurd to think
that the leaves are to be sustained by what they cannot
possibly obtain.

While the trees are in nursery rows, they will require
pruning in order to give them the proper shape when
wanted for permanent plantations. It is not necessary to
prune severely, but just enough to give the growth the
proper direction. If intended for timber trees, then a
tall straight stem is required, and when there is more
than one leading shoot, they should be cut away. The
lower branches may be removed from time to time, always
leaving enough to form a good head to the tree, and in
cutting off branches, they should be severed close, leaving no rough stump to decay, or to throw out sprouts.
If the young trees are properly pruned in the nursery,
there will be no necessity of removing large branches
when they become old. Some species will require but
little pruning, while others demand considerable, else
they make slow progress in the way of making handsome
shapely specimens. Pruning should not be practised
to such an extent that the tree is weakened by the operation, but it should never be neglected when anything can
be gained in promoting the growth of any part of the
tree, or in any direction that will tend to increase its
value, or fit it for the purpose for which it is raised. By
cutting off a portion of the lower branches, we allow
more sap to flow past, and into those higher up on the
stem, and we repeat the operation annually, or as often
as necessary to encourage an upward growth if tall trees

are our object, instead of low and very stocky ones.
Trees growing in an open field and left to themselves, will
usually have branches sufficient to shade their stems.
This appears to be not only natural, but beneficial, for
when the stem is fully exposed to the sun, the bark becomes dry and hot, and the flow of sap is retarded in its
movements. It is only, however, while the trees are
young and the bark thin, that any particular injury will
be perceived.

When the trees are raised in nurseries, the stems are
partially shaded ; consequently the lower branches are
not required for shade, but only to assist growth until a
sufficient number of others have been produced, and then
their service may be dispensed with without injury to
the tree.

Trees standing singly and alone where they have room
for full development, should have at least two-thirds of
their hight occupied with branches, but where grown in
forests for timber, the rule may be reversed, although we
may vary the proportion of occupied and naked stem, according to the natural habit of the tree. The pruning
of forest trees should not cease with their final planting
in the position in which they are to remain, because an
occasional lopping off of a branch here and there, removal
of sprouts from near their base, or suckers springing
from roots, may assist greatly in keeping them in good
shape, and prevent the growth of parts not desired.
Stunted, distorted specimens, may often be entirely ren-
ovated, as it were, by judicious pruning.

TIME TO PRUNE.

This is a subject which has been frequently discussed
among arboriculturists, and all who cultivate trees of
any kind, but all will agree that it should never be done
at a time when the sap will flow from the wound, as this
not only causes a loss to the tree, but the slowly oozing

sap has a corrosive action on both the exposed wound, and surrounding bark, often hastening decay. This is especially true with trees like the maple, butternut, and birch, which bleed (as it is termed), if wounded at any time during the latter part of winter or early spring. The oozing sap also attracts certain insects, especially those that infest dying or dead wood. In my own experience I have never found any better time to prune than in summer, as soon as the trees are in full leaf, and the trees have commenced to make a new growth. The wounds made at this time will commence to heal over immediately, and where small branches are removed on rapidly growing trees, the wounds will usually be entirely covered with new wood by the end of the season, and where larger branches are cut off, the exposed wood will become well seasoned, and so hardened during the warm weather, that it will seldom commence to decay before it is entirely overgrown. The next best season is in the fall after the wood is ripe, for in cool climates the exposed wood will become dry, and hardened before the sap commences to flow in spring.

PRUNING EVERGREENS.

The conifers and other evergreens will submit to the knife and the pruning saw, as well as deciduous trees, and when raised for timber, will need pruning as often, and in about the same manner. When raised for ornamental purposes, the pruning will be mainly for the purpose of giving them the required form, although thinning out, and shortening the branches at the time of transplanting, is as beneficial as it is with deciduous trees, but it it is not so generally practised. Evergreens may be headed back or trimmed up in order to make them grow tall and slender, or broad and stocky. With the natural conical shaped evergreens, like the spruces and balsams, many persons dislike to cut out the leading shoots, for fear of destroying the natural symmetry of the tree, and while it may have this effect for a short time, a new leader is certain to come in and take the place of the one removed, but during the time intervening, the lateral branches will spread out more vigorously, giving to the tree a more stocky appearance. In pruning the coarser growing pines, a little more care is required than with arbor-vitæ, spruces, and other closely allied trees, for the reason that buds are not usually produced on the internodes between the nodes or joints, and when a leading shoot, either the terminal one on the main stem, or branches, is removed, it should be cut out close down to the junction of the next tier of branches below, leaving no barren stump to die and decay. A glance at a pine tree will be enough for even a novice in such matters, to see how it should be pruned, in order to make it grow more compact and stocky, if such a change is desired.

IMPLEMENTS USED IN PRUNING.

The common pruning knife is the best implement for pruning small trees, but in removing large branches, a fine-

Fig. 24.—A HANDY LADDER.

tooth saw should be used in preference to an axe. If the wounds made are so large that they will not soon be covered with a new growth, it is well to apply some kind of wax, paint or some other substance, to exclude water and prevent decay. Various compositions are used for this purpose, and on small trees where the exposed wood can be readily reached, a little melted grafting-wax, ap-

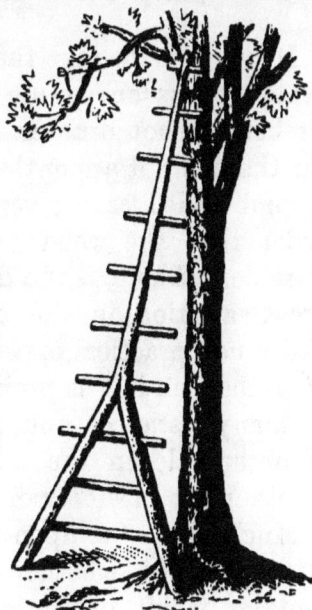

plied with a brush, will be found an excellent preservative, but on large trees where there is considerable surface to be covered, almost any good mineral paint mixed with linseed oil will answer every purpose. A handy and cheap ladder for forest-tree pruning is shown in figure 24.

CHAPTER XI.

THE BEST TIME TO CUT TIMBER.

If we were to take the opinions of men, practical and otherwise, as our guide in selecting a time for cutting timber we should never reach a conclusion in the matter, for there is not a month in the twelve, that has not been recommended as the very best time for felling trees in order that the wood should remain sound, firm, and durable. There is, no doubt, some foundation for this great variation in the opinion of even those who have had much practical experience in handling and working of timber, and it is probably largely due to the fact that in many instances, and for many purposes, no difference is observable in the appearance or quality of timber whether cut in winter or summer.

Much depends upon the treatment timber receives after it is cut, whether placed in a position to season rapidly, or left in the woods where seasoning will go on slowly; furthermore, climate—the prevalence of insects that attack felled trees—the kinds of timber, and various other conditions and circumstances has much influence on the durability and quality of wood of the same species of trees. It is certainly true that there is a great difference in the amount of, and condition of the moisture in trees at different seasons of the year, and while as a matter of convenience it will often be of more importance to the one cutting timber than any slight variation in quality that may follow, still there is no doubt a choice in time for felling trees for all purposes. In late fall and winter, when trees are in a dormant state, the wood contains less liquids than in spring and summer, and this is without doubt an advantage, for there is not only less to be driven off in seasoning, but less to produce chemical changes which are often more or less injurious to both strength and durability.

From my own experience and all the facts that I have been able to gather from lumbermen and dealers in timber, I have come to the conclusion that the winter is not only preferable but the most convenient season for cutting timber, whether to be converted into sawed timber or be used for posts, rails, railroad ties, or other purposes where toughness and durability are an object. But in case of small timber for posts and stakes from which the bark is to be stripped, then we may delay the cutting until the latter part of the winter, or until the sap commences to liquify, which will facilitate the removal of the bark. All stakes and posts which are to be set in the ground should have the bark removed, certainly on that portion which is placed in the ground.

If we bear in mind the fact, that it is only the outer portion of the tree—the sap wood, leaves, buds, and inner layers of the bark—which are alive and contain true sap, all other portions being dead, and only serve as a covering, or like the heart wood, help to sustain the tree in its position, we can readily see why it will make no material difference in the lasting properties of timber whether it is cut in summer or winter, provided the green portion is soon deprived of its moisture, so that insects will not find a lodgment for their eggs or decay be accelerated by its presence.

For such purposes as hoop poles, the bark must be retained as it is generally considered essential, and in

4

this case the young trees should be cut at a season when the bark will adhere the most firmly, that is late fall or early winter, although they may be cut in summer, because the hickories usually finish their growth quite early in the season, but the wood is likely to be more brittle if the poles are cut early or when in full leaf than later.

Coniferæ trees from which it is desirable to strip the bark should be cut during the growing season in early summer, and if rapid seasoning is desired without removal of bark, the trees should be merely felled and allowed to remain with all their branches attached until the leaves fall off.

We may have other objects in view besides the value of the timber taken, such as a second growth to be produced from the stumps, when this is desired the trees should be felled at a season most favorable to the roots. If the trees are cut late in the fall or winter, the roots and stumps will throw up sprouts far more readily than if the trees were cut in summer when growing the most rapidly. In fact, late summer is the proper time to cut trees and shrubs if we desire to kill the roots. It would not be possible to name the exact time best for the purpose, because not only do seasons vary but the right time in New York State would be too late for Virginia and those further South, neither is the same time best in all years. I have seen acres of willows killed out completely by a single cutting of their tops, and the next season another lot was cut off during the same days of the same month, the roots of which were but little injured and threw up sprouts in great abundance the following season. The weather at the time of cutting the willows, no doubt, had some influence in producing the difference in the results noted.

In ancient times, and, in fact, in modern, many persons have believed the moon has some mysterious influence upon the growth of animals and plants inhabiting

this earth, such insist that trees should always be cut
during certain phases of the "pale orb of night," but for
some reason they fail to agree in this matter, some in-
sisting on the wane, others the new, etc., but such super-
stitions have long since become obsolete among men who
know anything of natural history in any of its various
branches.

CHAPTER XII.

IMPORTANCE OF A SUPPLY OF WOOD.

No one who is at all familiar with forests and their
products, needs to be reminded of the importance of
having at hand an abundance of wood of various kinds,
or how much it contributes to the general welfare and
happiness of a nation. But there are those who have
not paid much attention to this subject who claim, and
no doubt honestly believe that the great progress made
of late years in the use of iron in place of wood in building houses, bridges, piers, ships, and other structures,
are but indications of what is to follow, and that in a
few years there will be no great demand for wood in any
form.

The building of railroads, which reach almost every
part of the country, has aided in the distribution of
coal, and made this in a great measure a more convenient
and in many instances a cheaper fuel than wood, but in
building these roads a vast quantity of wood has been
used, and of the best kinds, not only for ties, of which
nearly or quite three thousand are put down per mile,
but on many of the roads wood is still used for fuel.
There is now nearly or quite one hundred thousand miles
of railroads in the United States, and we have only to

multiply this by three thousand, to ascertain that three hundred millions of ties have been used in their construction, leaving out of account the thousands of wooden bridges and other structures, in the building of which more or less wood has been consumed. The railroads may have assisted very materially in checking the consumption of wood for fuel, but they have probably more than balanced the account in the amount used in their construction, besides the three hundred million of ties must be duplicated every ten years, for the average life of a railroad tie will scarcely exceed a decade, and with nearly all kinds except the best oak, it is a year or two less.

.The demand for railroad ties is not likely to decrease, but increase, although as timber becomes scarce and prices advance, preserving processes will doubtless be employed to prevent rapid decay. Stone, brick, and iron will also come into more general use for buildings, but the increase in population will also tend to an increase in the demand for other purposes besides that of buildings.

It is only a little more than a century since coke was first employed for smelting iron ores. The introduction of this fuel to take the place of charcoal, it was thought would save the forests of the world from destruction by the charcoal burners, and while it has done much towards making it possible to produce sufficient iron to meet the great and constantly increasing demand, it has not superseded charcoal, and there is probably more charcoal used to-day than at the time coke was first employed in a smelting furnace. Charcoal is still used in furnaces and forges, and there are several establishments in this country that use annually over a million of bushels each, and a score of others that consume from twenty to twenty-five hundred thousand bushels.

Notwithstanding the number of substitutes that are employed, the demand and consumption of wood appears to increase, and to-day there is probably more wood used in making boxes of various kinds than there was in the construction of buildings of all kinds in this country three-quarters of a century ago. Furthermore, no kind, or quality of timber appears to escape the unsatiate demand of the artisan of the period, and he not only finds ready uses for the large and small, the hardest, toughest, and most durable, but also for the soft and spongy, the latter being preferred for grinding up into wood-pulp for making paper.

Not a year passes during which scores of new devices and inventions of new articles of manufacture, are not brought forward, that are made in part or wholly of wood, and while singly they may not call for a great quantity, they do in the aggregate use up an enormous amount. '

The invention of a pleasing toy for children has frequently caused the demolishing of hundreds of acres of forests, to supply the manufacture with wood used in its construction. It is idle to talk of our natural forests furnishing a supply of wood for the future use of our people, even with the most careful management and economy in preventing waste, there must soon come a time of great scarcity of all kinds of wood. With an increase in population, there must necessarily follow a corresponding increased demand, because experience has shown, that whenever any other material has been substituted for wood, it merely releases a certain amount, and allows it to seek other channels or markets. No matter in what direction we turn, the fact meets us, that the best and most valuable forests of the United States are rapidly disappearing, and the sooner we commence as a nation to economize in the use of wood of all kinds, and preserve the forests now existing, as well as commence planting new ones, the better it will be for the present as well as future generation.

It is not necessary to select the best and most fertile land upon which to raise trees, for any that is rich enough to give the plants a good start in life will answer, because the annual dressing of leaves that the soil receives will be sufficient to keep the trees growing. There are doubtless many situations, where a single tree would not thrive, as on a prairie, a bleak hillside, or other exposed positions, where by planting a number together they would mutually protect each other, and will usually take care of themselves. We have millions of acres of barren, naked, sandy, rocky, and otherwise unproductive lands, that might readily be covered with valuable forests. Large plantations of forest trees have been established in Europe, and there is no good reason why the same should not be done in America.

CHAPTER XIII.

PRESERVATION OF FORESTS.

In the first settlement of our Atlantic Coast there was an actual necessity for clearing off the forests, in order to obtain land for cultivation, and while at this day the greater part of our arable lands has been cleared, there is still quite large areas well adapted to cultivation and awaiting the husbandman. But there are still larger areas of hills and mountains that are not, and probably never will be worth clearing for any agricultural use, and as such lands are to a large extent still covered with forests, it is not too late to attempt their preservation. These wood-lands have, it is true, been overrun more or less and the best timber removed, but this has not to any great extent affected their value in the way of influence on the climate of the surrounding country, and as sources of water supply to feed our brooks and rivers.

The Adirondack region of the northern part of the

State of New York is one to which public attention has been called of late, and while the importance of preserving the forests over this entire region of country can scarcely be questioned, it is at the same time only one of many similar areas that should become public domain and the forests covering them remain inviolate for all time. There is scarcely a brook or river, from the Atlantic to the Pacific Ocean, that does not flow from some forest-covered hill or mountain, and this in itself is enough to warrant the withdrawing of all such land from market, whether owned by the different States or the General Government. There are large areas covered by forests throughout the entire Alleghany range of mountain, from Pennsylvania to Georgia, also in the Blue Ridge and the Cumberland Mountains, all of which should be preserved as public domain, instead of being disposed of by the different States in which they are located for a few cents an acre, as has been done in thousands of instances.

Similar areas of wood-lands, but less in extent, may be met with in the same latitude, until we reach the Pacific slope. But the best and most valuable timber in all of these forests is being removed at a rapid rate, and if it is to be preserved, no time should be wasted by the different States in which they are situated, or the General Government, in taking possession of them.

Laws may have to be enacted looking toward the control and general management of these forests, and schools of forestry established, where young men may obtain the information required to fit them for the position of foresters, but these are trifling matters in comparison with the more important one of securing and establishing State and National forests.

MANAGEMENT OF FORESTS.

I have no doubt that many persons will object to this proposition of passing over large areas of forests to the control of the States or General Government, on account of the expense likely to be incurred in their purchase and management. But it is not at all probable that these forests will become a burden to the people; but on the contrary, if properly managed, may be self-supporting, if nothing more. With proper management there should be an income from the sale of timber of various kinds, for when a tree has reached maturity, it ought to be removed, else a decrease in value will ensue. Skillful foresters will not only remove and dispose of valuable timber at the proper time, but be constantly planting trees in all available grounds throughout the forests under their care. The inferior kinds will be removed to give room for the superior, and in this way the forests may be improved and their intrinsic value enhanced very materially from year to year.

What kinds should be destroyed as well as planted, depends so much upon soil, situation, climate, and local demand, that no general rules can be given for such operations, but must be left to the good judgment of the forester himself, or his counsellors. They will also be best able to decide whether it is better to plant young trees or sow seeds, where a new growth of wood is desired.

CHAPTER XIV.

ESTABLISHING NEW FORESTS.

In the great treeless regions of the West—forests must be raised if any are ever to adorn that part of our country, but there are extensive areas on which it will be extremely difficult to make trees of any kind grow without irrigation, and to do this some heretofore undiscovered source from which a supply of water can be obtained must be brought to light. But we may well leave the higher and drier regions west of the one hundredth meridian for future generations to experiment upon, for the present has enough to do in raising forests on more congenial soils. There are limited areas where both soil and climate are so well adapted to the growth of trees that forests can be started by merely scattering the seed over the ground, and leaving them to sprout and grow without further care or attention, but while this system may answer for such kinds as locust, maples, and elms, and on soils quite free from rank growing grasses and weeds, other kinds would fail unless covered with earth, or at least shaded until they had produced roots and become fixed to the soil. When trees are started in this way there can be no uniformity in their distribution, and while some will be crowded others will have more room than is necessary ; consequently, if anything like system or regularity is to be secured, there will need to be more or less thining out and transplanting done, and this will cost nearly or quite as much as it would to have sown the seed in nursery rows and then transplanted the seedlings when of proper size and age. Sowing forest tree seeds on unbroken soils on the banks of rivers and smaller streams, or in forests where the trees are very scattering. has often been practised with excellent results, and is to be recommended for those who cannot afford to adopt a more advanced system of tree culture. Such half-wild plantations will also furnish trees for trans- planting to other locations if they are needed, but the usual system of raising seedlings in beds or nursery rows, will, as a rule, give the best and most satisfactory results.

The first thought of the pioneer in a forest covered region is to clear off the trees—let in air and sunlight in order that the earth may be warmed, dried, and fitted for cultivation and production of such crops as are re- quired for the maintenance of man and his domesticated animals. His crops also need to be stirred by gentle breezes to keep them in health and insure vigorous growth, for a stagnant atmosphere is no more to be desired than stagnant pools of water, but unfortunately, man in his anxiety to secure a large area of land for cultivation, allows his greed of gain to get the better of his judgment, and the onslaught on the forests continues until there is no shade or protection from the hot rays of the sun that parch and dry up his fields, and instead of opening the way to gentle life-giving breezes, he has admitted the fierce winds and tornadoes. On the contrary the pioneers on the plains and prairies, need, not only protection from the fierce rays of the sun in summer, but against the winds which sweep over those regions with a violence and frequency only known to those who have encountered them at all seasons and in all kinds of weather.

How to begin in order to get trees growing in such numbers as will afford shelter and protection, is what most interests those who have resolved to make themselves a home on the prairie. They are not, as a rule, very particular as to the kind planted, because a tree of any species is so much gain, and a thing to be admired, appreciated, and tenderly cared for. Admitting that every tree raised is a gain, and a step towards securing what is so generally sought by those residing in sparsely wooded regions, it may be well, at the same time, to take into consideration in advance of planting, not only the present, but also the future value of the kinds to be employed in forming screens, wind-breaks, or even more extensive plantations. The poplars and willows have been most extensively planted, probably, because they could be easily obtained and readily propagated by cuttings. They also grow rapidly even under what may be termed unfavorable conditions, but the wood is very inferior, and while it is better than none, it does not answer the same purpose as that of many superior kinds that can be readily produced in the same region of country, and under the same natural conditions. Some of the species of poplar known as the cotton-woods have been extensively planted in the States west of the Mississippi, and while the trees grow rapidly in hight, they do not spread out and assume a sturdy, stocky habit, such as is needed to effectually resist the force of prevailing winds. If planted thinly and each tree given abundant room with an occasional cutting back of the leading shoots, they would serve the purpose better, but as I have seen them planted in hundreds of instances, with no thinning out or heading back, the plantation in a few years had more the appearance of a collection of hop-poles than anything else, and usually they lean over to the east, or south-east, at such an acute angle that there is no mistaking the point of compass from which the wind blows most persistently if not continuously in those regions.

While the poplars and willows have, no doubt, served a good purpose, and may still be employed for screens and timber belts to a limited extent, they ought never to be recommended for anything more than temporary plantations, or to foster better kinds. There can be no reasonable excuse in these days for planting inferior kinds of trees, because it costs really no more to raise the best from seed—dig up from the woods, or procured from the nurseries—than it does to handle or purchase the poorest. In all cases I would advise planting young trees or cuttings in ground that had been broken up at least one year before being used, and the planting in all cold climates should be done in spring. The more carefully the ground is prepared for the reception of the plants the better, and the strongest and most hardy should be placed in such a position that they will protect the weaker and tenderer kinds. Each species of tree, as a rule, should be kept separate and not intermingled as they are often found in a state of nature. Of course many kinds and varieties may be employed in forming belts, groups, or forests, and still each be placed in separate rows, squares, or clumps, but this system may be varied in case some small and less vigorous species are needed to fill in among the larger ones in order to give . compactness to a plantation intended mainly as a screen or wind-break.

The object in keeping each species separate is to avoid giving any one an advantage over its neighbor, which is certain to follow intermingling of different species. It may answer in some cases to intermingle several different species of the oak, maple, and similar trees, still, we seldom find that the different species of oak or maple, do equally well on the same kind of soil, and for this reason it is best to keep them separate in our cultivated planta- tions, in order that we may the more readily determine which is best adapted to the soil and climate. Evergreen trees are superior to deciduous for screens and wind- breaks, but more difficult to raise on the prairies because of the exposure of their leaves to drying cold winds in winter. But by selecting those species that are indig- enous to similar soils and climates, and then by giving protection in winter until the trees become well estab- lished, I am inclined to believe that a very fair variety of evergreen trees may be made to thrive in almost any locality where deciduous trees will grow. It will not be necessary to obtain trees from extreme northern latitudes in order to find species that will succeed in Minnesota, Nebraska, or further South or West, because it is not so much the low temperature that destroys them, as it is exposure to cold drying winds. For instance, in my grounds I have three large American hollies planted some fifteen years ago, one of these trees is protected on the north-west side by a small clump of American arbor-vitæs, and it has never been injured in the least by cold, and is every winter loaded with its bright, scarlet berries. The two other trees, not more than a hundred feet distant, but unprotected, are frequently badly injured and occasionally lose all the leaves from the north side. Now the injury to the two unprotected specimens cannot be attributed to the difference in temperature, for if a thermometer was hung up in each they would not show a difference in temperature of a single degree even in the coldest weather. It is the cutting wind that kills, and not the severe cold. Evergreens that are indigenous to cold, moist climates, will not thrive in cold, dry ones, neither in those that are moist and warm. Our northern species, like the hemlock, white spruce, white pine, and arbor-vitæ, will not grow in the Southen States, except on some mountain range where the temperature is not excessive in summer. There are, however, a good variety of evergreens indigenous to the Southern States, and adapted to all kinds of soil from the dry sand hills where the long-leaved or yellow pine flourishes, to the low swamps filled with white cedars, evergreens, oaks, yews, and magnolias. But in seeking evergreen trees for cultivation on the western prairies, it will be well to obtain species inhabiting similar parallels of latitude, and those known to resist high winds and long drouths. Such species can be found in both the Eastern as well as the Western States.

Among the pines, those with coarse rigid leaves are less liable to be affected by strong winds than the more soft and tender-leaved. The common Pitch pine (*P. rigida*), and the Jersey pine (*P. inops*), as well as the Table Mountain pine (*P. pungens*), and Red pine (*P. resinosa*), are species well adapted for planting in exposed situations. There are also several species, natives of the foot-hills and mountains, bordering the great plains on the west that will eventually prove to be more

valuable for planting on the prairies than any of our
Eastern species, and among them I would recommend
the Heavy-wooded pine (*P. ponderosa*), because it seems
to be almost indifferent as to soil and location, and I
have seen it growing luxuriantly in the most exposed
situations in the mountains, among rocks where there
was little or no soil—in the hardest clay as well as in
loose beds of gravel, and this too in regions where thirty
degrees below zero in winter is not an uncommon temperature.   I refer to this pine as one likely to succeed
under the most adverse conditions, but there are other
native species probably more desirable an account of
appearance as well as quality of timber, but we are
now seeking trees that will resist winds, drouths, and
give the pioneer on the prairie something to cling to,
after which the more beautiful and useful among trees
may receive attention.

There are also several other species of pines, spruces,
and a red cedar, found in the same regions along with the
Heavy-wooded pine, all of which are worth trying as they
may succeed perfectly, but in all cases I would advise
obtaining seeds or plants from the higher and colder
parts of the mountain for planting on the prairies, because those from the warmer and moist valleys would
suffer more from the change than those from drier and
more exposed positions.   There are also several foreign
species of pines and other cone-bearing trees that would
probably succeed as well, or nearly so, on the prairies as
those I have named, but of these I shall have more to
say hereafter.

Those who are about beginning to establish forests, or
even limited plantations, should carefully consider the
adaptation of trees, not only to climate but soil, for some
species succeed only in moist or wet soils, others in dry,
while a few may appear to do equally well in both.
Then again certain species only thrive on sand stone forma-

tions, or where slate or granites predominate, and utterly fail on limestone, or what are termed rich limestone soils, which some claim to be the case with the chestnut. Some trees appear to require opposition or resistance to root growth in order to keep them healthy, and these kinds do best in stiff clay or on soils filled with loose rocks and similar obstructions. We can usually make a very close guess as to what kind of soil is best adapted to a species if we know the character of that in which it is naturally found most abundant, and for this reason we would not select a clayey soil for the white pine, or a light sandy one for the elm, hickory, or maple. Then again we would much prefer a swamp for the red maple, and a hard, dry and moderately dry soil for the sugar maple. I throw out these hints in order that those who may have occasion to make selections from the trees described in the following pages, will not overlook whatever I may have to say in regard to their native habitats.

CHAPTER XV.

FOREST TREES.

I propose in the following pages to mention all trees indigenous to the United States, so far as known to botanists, also the best known of the exotic species that have been introduced and cultivated to any considerable extent for ornamental or other purposes, but as the limits of this work will not admit of a full botanical description of all the species and varieties, I shall only refer to some of the most conspicuous and familiar characteristics of each, and in language that I hope can be understood by those who are not accustomed to the use of purely scientific terms. Those who may desire a full scientific description of the trees mentioned in these pages, can find it in various botanical works published in this country and Europe.

Shrubs that seldom reach the hight of twenty feet are omitted, except in some instances where they belong to a genera containing trees of larger growth, and in such instances they will be mentioned briefly. I have arranged the list of trees alphabetically according to their generic name, and while this is not in accordance with the botanical classification, it will be found just as convenient for all practical purposes. I make only two classes, the first comprises the deciduous trees and broadleaved evergreens which are principally indigenous to the Southern States, and the second the conifers or cone-bearing trees, the greater part being evergreens. There are a few like the Larches, Taxodiums, and Salisburias that are deciduous trees, but they belong among the true conifers.

ACACIA GREGGII, Gray.

A small tree, but sometimes over twenty feet high. Leaves small, short, composed of two or three pairs of pinnæ an inch long, and leaflets of four or five pairs, oblong or oblong-ovate. Flowers in cylindical spikes an inch or two long, succeeded by curved pods three or four inches long. Seed about a half inch long. Branches either naked or armed with stout-hooked prickles. Wood firm and hard, but usually too small to be of much value. Native of Texas and westward to Southern California. There are quite a number of the species of the Acacia that have been introduced from tropical countries, and are now naturalized in the Southern States, also a much larger number that are cultivated as green-house plants.

ACER.—Maple.

An extensive genus, containing some fifty species, mostly of the northern hemisphere, and pretty evenly distributed through the northern border of the temperate zone in America, Europe, Asia, and Japan. There are nine or ten species, natives of the United States, more than half of which are valuable timber trees. Our native species have palmated-lobed leaves, with edges variously toothed or notched. Flowers small in terminal racemes or umbel-like corymbs.

Acer Saccharinum.—Sugar Maple, Rock Maple, Hard Maple.— Leaves three to five-lobed, deep green above, and paler beneath. Flowers greenish-yellow, appearing with leaves in spring. Wings of seed quite broad, seed ripe in autumn. A well-known tree of rapid growth, possessing many valuable qualities, one of which is its sweet sap, from which large quantities of sugar are made in regions where the tree is abundant. The wood is hard, close grained, and susceptible of a fine polish, and extensively use for hard-wood floors and inside finishing of houses, also, for cabinet work, especially what are termed "Bird's-eye" and Curled Maple. Hard maple makes an excellent fuel, and is highly valued for this purpose. A rapid growing tree often reaching a hight of eighty to ninety feet, with a stem three to four feet in diameter. Most common in the North, from Maine to Minnesota, and also southward to Georgia in the mountains. Succeeds best in rather strong, loamy soils, approaching a stiff clay, and on stony hill-sides and ridges where the soil is moist, but not wet and swampy. A variety of the Sugar Maple found in some of our Northern woods called the Black Maple, has darker green leaves which appear a few days later in the spring than this species. The Sugar Maple has long been a favorite for planting in the streets of our cities and villages, also as a roadside tree in the country. It is well worthy of all the attention it has received, and should be more extensively planted wherever forest trees of any kind are needed. It is so abundant in the Northern woods that seedlings of almost any convenient size for transplanting can be obtained in unlimited quantities, and at a mere nominal price of those who make a business of gathering them for sale.

A. dasycarpum.—White Maple, Silver Maple.—Leaves deeply five-lobed, silvery white underneath ; pale green above, lobes coarsely cut and toothed. Flowers greenish yellow or reddish without petals, appearing in early spring, succeeded by the corymbs of winged seed, which are ripe about the time the leaves are of full size. The seeds soon drop off, and where they fall on moist soil in the shade they soon grow. They are very delicate, however, and cannot be kept for many weeks after they are ripe, but if sown immediatedly and in good soil they will produce plants two feet or more in hight the first

season. This is one of the most rapid growing of all our
maples, and succeeds in a great variety of soils, but is best
adapted to a rich, moist one. Its wood is white, fine grained,
and rather light and soft, but takes a fair polish, and is much
used for purposes where a very hard surface is not required. The
sap is sweet, and sugar can be made from it, but is much
inferior to that of the Sugar Maple. Occasionally a tree yields
the accidental form known as Curled and Bird's-eye Maple.
This species of maple has been raised in large quantities by
Eastern nurserymen, and sold for planting in streets and parks,
for its rapid growth and adaptation to almost all kinds of soil
and situation, has made it a general favorite with those who
desire to secure shade trees with as little delay as possible. The
tree in favorable soils often reaches a hight of eighty feet or
more, with stem three or four feet in diameter. I have raised
trees from seed that were ten feet high at the close of the fourth
season, and in twenty-five years, more than forty feet high,
with stems eighteen inches in diameter at the base, and this too,
in rather light and only moderately rich soil. The White
Maple is more abundant west than east of the Alleghany
Mountains, although it is found sparingly in Northern Vermont, and thence westward to Minnesota, and southward to
Florida. When planted singly it forms a large spreading top,
the outer branches often becoming somewhat pendulous or
drooping. While we have many better timber trees than this
species of maple, still its rapid growth and adataption to such a
great variety of soils, and wide range of climate, gives it a
value possessed by no other species, and it deserves more attention than it has ever received from those who are in haste
to obtain shelter and good fuel in a few years, and with little
expense. The branches are abundant and flexible, a merit of no
small moment with trees to be employed as wind-breaks in prairie
regions of country. There are several ornamental varieties of
this species cultivated in nurseries, among which the following
are desirable as lawn trees, or for planting in parks, and other
pleasure grounds : Crisp-leaved (*A. dasycarpum*, var. *crispum*);
leaves deeply cut and much curled ; more or less upright.
Wagner's Cut-leaved (*A. d. Wagnerii laciniatum*), a handsome
variety with divided or cut leaves. Weir's Cut-leaved (*A. d.
Weirii laciniatum*), a very graceful tree usually of weeping habit,
but in some specimens the branches assume a wide, spreading
habit, and droop but slightly or not at all. Varieties of the

White Maple may be readily propagated by budding or grafting upon seedling stocks of the species.

A. rubrum.—Red Maple, Scarlet Maple, Swamp Maple.—Leaves usually three-lobed, as shown in fig. 25, but sometimes five, the middle one the longest, all irregularly serrate. Flowers crimson-scarlet, and sometimes yellowish, appearing early in spring, succeeded by smooth seeds with spreading wings, about an inch long. Seeds ripen early, or by the time the leaves have fully expanded, and then drop off and soon decay, unless placed in a favorable position for growth. Wood white, or slightly tinted with red, close-grained, and moderately fine; a little heavier than that of the White Maple, and more extensively employed for cabinet-making and various articles of wooden ware. Valuable for fuel, but not equal to the Sugar Maple. This species also furnishes Curled and Bird's-eye Maple for cabinet work. A very large tree, and common in nearly all swamps in the Eastern States, and sparingly in the Western, also occasionally found as far south as Florida. When planted singly, it forms

Fig. 25.—LEAF OF RED MAPLE.

a handsome round-headed tree, not as open and spreading as the Silver Maple, neither is it of as rapid growth, but with age it reaches fully as large a size. Although naturally found in swamps, the Red Maple will thrive in moderately dry soils, and is often planted along roadsides, in preference to other species, on account of its brilliant-colored flowers in spring, and the various colors of the foliage in autumn. The coloring of the leaves of this species is a puzzle to the scientific naturalist, for there appears to be no accounting for the many colors, or

their distribution, not only among different trees growing under
exactly the same conditions, but on different parts of the same
tree. Sometimes the leaves on a single branch will change to
an intense crimson or scarlet, while those on other branches will
retain their normal color until cut by frosts. Then, again, one
tree in a row will assume the scarlet or crimson color, and those
adjoining will show very little, if any coloring, except perhaps
a faded red or yellow; but the very next season these colors
may be reversed.

The Red Maple is not only a handsome tree, but well worth
cultivating, both for ornamental and useful purposes. There
are several varieties in cultivation, but not sufficiently distinct
as to have attracted much attention. *Acer rubrum fulgens* is a
dwarf variety, and *A. r. globosum* is a variety with a globose, or
round head, while *A. r. pyramidalis* is a very distinct pyramidal form.

A. Spicatum.—Mountain Maple.—Leaves slightly three-lobed;
coarsely toothed ; downy beneath, with dense, upright racemes
of flowers appearing very late in the spring, succeeded by small
seeds with narrow wings. It is only a small shrub, six to ten
feet high, found in the Northern Border States and on some of
the higher mountains southward. .

A. Pennsylvanicum.—Striped-bark Maple, Moose-wood, Striped
Dog-wood.—Leaves large, thin, somewhat heart-shaped, but
with three-pointed, serrated lobes. Flowers greenish, in terminal racemes, appearing after the leaves. Seeds with large, divergent wings. A small tree, with light-green bark, striped with
darker lines. Sometimes cultivated as an ornamental shrub or
small tree.

A. circinatum.—Round-leaved, or Vine Maple.—Leaves rounded;
seven to nine lobes ; serrate. Flowers purplish, in small clusters. The wings of the seed diverging in a straight line. A tall
shrub, but in some situations reaching a hight of thirty to forty
feet. A native of Northern California, and northward to
British Columbia. Wood very hard and fine-grained, but not
plentiful enough of large size to be worthy of much attention.

A. macrophyllum. — Large-leaved Maple, California Maple.—
Leaves very large, deeply five to seven-lobed, with very coarse
teeth. Flowers of a yellowish color, in a compact raceme.
Fruit hairy, with large, broad wings. A very large tree,
sometimes one hundred feet high, with stem five feet or more

in diameter, but only on very favorable situations does it
grow to such a size. Wood very hard, resembling that of the
Sugar Maple, and one of the best and most valuable hard woods
found west of the Rocky Mountains. The sap is sweet, and
yields a fair quality of sugar. This maple occurs in California,
from Santa Barbara, and northward, to Washington Territory.
It is a tree well worthy of the attention of arboriculturists,
East as well as in the West ; but the seed should be procured
from Northern localities, and from large trees, else the plants
are likely to be tender and of slow growth in localities east of
the mountains.

A. grandidentatum.—Mountain Sugar Maple.—Leaves slightly
cordate or truncate at the base, pubescent beneath, and rather
deeply three-lobed; lobes acute with a few sinuous indentations.
Flowers few ; the petals nodding. Seed smooth, with small,
diverging wings. This species, although closely related to the
Sugar Maple, does not attain a very large size, seldom growing
more than thirty feet high. It is found in Arizona, Southern
Utah, and on the west side of the Mountains, near the head-
waters of the Columbia, principally in the valleys, and near
small streams.

A. glabrum.—Smooth-leaved Mountain Maple.—Leaves smooth,
two to four inches broad, rounded, heart-shaped in outline,
with rather shallow indentations, although occasionally distinctly three-lobed ; the lobes doubly serrated, with acute teeth.
Flower in large corymbs, on short branchlets ; greenish-yellow.
Seeds, with broad-spreading wings, ripen late in fall. Quite a
variable species, both in leaves, color of the branches, and form
of growth. This species probably grows at a higher elevation
in the Rocky Mountains than any other native maple. I have
found it abundant in Colorado and New Mexico, at an elevation
of ten thousand feet. In exposed situations, on the sides of a
canyon, it was merely a tall shrub, with many stems springing
from the same root, probably because frequently killed down
in winter ; but where protected by other trees, it assumes an
upright form, growing thirty or more feet high. Wood quite
hard, and fine-grained, but, as generally found, it is too small
for any practical use except for firewood. Common in the
mountains of Northern New Mexico, Colorado, and west to the
Sierra Nevada, and northward to Vancouver's Island.

A. Negundo, or Negundo aceroides.—Negundo Maple, Box Elder,
Ash-leaved Maple.—Although our modern botanists consider

this tree sufficiently distinct to be separated from the true
maples, it is however so closely allied to them, that for convenience's sake I have named it here. *Negundo aceroides* is the generic name most generally employed in botanical works of the
present day. The pistillate and staminate flowers are produced
on different trees; consequently, in order to raise fertile seeds,
both sexes must be present, or the trees not far distant. Leaves,
pinnately three to five-foliate, the leaflets ovate or oblong, either
lobed or toothed. Flowers small ; greenish ; the fertile ones in
racemes from lateral buds, and appearing with or before the
leaves. The seeds are oblong, extending about half the length
of the wing, ripening in late summer or autumn. Wood moderately fine, white, and makes good fuel when well seasoned.
A tree thirty to sixty feet high and two feet or more in diameter.
A widely-distributed species, being found in Vermont, and
westward to Utah, and southward in the canyons of New Mexico
and Arizona ; also in Florida and Texas. A very hardy tree, and
has been planted quite extensively in Minnesota, and the colder
region of the Northwest. It is a very rapid grower while young,
but does not continue and become so large a tree as some other
species of Maple already named.   The California Box Elder
(*Negundo Californicum*) resembles the Eastern species very
closely, and was previously considered to be identical, but may
be distinguished by its smaller and narrower leaflets, which are
coarsely toothed, but less distinctly lobed.

There is a species of the Negundo indigenous to Mexico and
another to Japan, making four known to botanists. Varieties
occur among them all. but those in cultivation in this country
are of our native species. One of the most showy of these is
the Variegated Negundo, the leaves being distinctly marked
with white, but the tree is rather delicate and often kills down
in winter, still an occasional specimen will escape injury for
many years. There is one specimen at Rye, Westchester,
County, N. Y., now over twenty years old, that has never been
injured by the cold of winter or burning sun of summer.

The Crisp-leaved Negundo is another distinct and interesting
cut-leaved variety, and another known as *Violacea*, so named
on account of the peculiar color of the bark on the young
branches.   This last is a very vigorous-growing tree, and the
young shoots rather larger than those of the species. A pistillate tree of this variety, twenty years old, in my grounds fruits
heavily every year, but there being no staminate tree of either

the species, or any of the various varieties within several miles of it, the seeds produced are false. I have purposely kept this tree isolated from the other sex of the same species in order to see if by chance the flowers would be fertilized by some of the other species by which it is surrounded, for in that case a hybrid might be produced, but thus far nothing of the kind has occurred, and the seeds of the Negundo have been uniformly unfertile.

FOREIGN SPECIES OF THE MAPLE.

There are no European or Asiatic species of the Maple that for general usefulness are superior to the best of our indigenous species. But there are a large number of species and varieties well worthy of cultivation for ornamental purposes, and a few may be considered as useful forest trees.

EUROPEAN MAPLES.

A. Pseudo-Platanus.—Sycamore Maple.—A very large tree with rather coarse spreading branches and deeply, five-lobed leaves, rather downy beneath, and long reddish petioles (leaf-stalks). The seeds are produced in long, pendulous, spreading racemes, not in clusters or corymbs as in the Sugar, White, and Scarlet Maples. The Sycamore Maple is a very vigorous and rapid grower, even superior in this respect to our Sugar Maple, but its branches are coarser and not so numerous, hence the trees, when planted in streets or as single specimens on lawns or in parks, appear to lack that fullness and grace of outline that are so characteristic of the Sugar Maple. In Europe, the tree grows to a great size, sometimes reaching a hundred feet high. The wood is hard, close grained and valuable for many purposes. Old trees planted in this country produce seed in great abundance, and are usually to be obtained of dealers very cheaply.   .

There are several very handsome and desirable varieties of the Sycamore Maple in cultivation. The following are the most distinct : The Golden-leaved has deep, yellow leaves, occasionally streaked or mottled. Purple-leaved, leaves purple underneath and dark green above—the leaf-stalks also purple or reddish, a handsome and vigorous growing tree. Three-colored or Tricolor, leaves curiously streaked with red, white and green. Silver Striped, leaves striped and streaked with white, a very distinct and handsome variety, especially in spring when the leaves first expand. Velvet-leaved, a curious variety, with

velvety green leaves, but of rather dwarfish habit. Worle's
Golden-leaved, leaves spotted with yellow. Leopold's Striped-
leaved, leaves streaked with green, yellow, and white. Doug-
las's Sycamore Maple, leaves quite small, pointed, and of a
uniform yellowish color.

A. plantanoides.—Norway Maple.—A large round-headed tree,
resembling in general appearance the Sugar Maple, but is a
slower grower, at least this has been my experience with it,
and I think most cultivators of it will agree with me on this
point, but Mr. F. J. Scott, in his notes on this species in "Suburban Home Grounds" says : "This species has a more vigorous growth than the Sugar Maple." From my experience
I should not expect a Norway seedling to reach more than one-
half the size of the Sugar Maple in the same number of years.
The leaves are larger and thicker than those of the Sugar Maple,
but of the same rich, green color. The young twigs and buds
are a little coarser, but the bark on the twigs, larger branches,
and stem of the trees is very similar in general appearance to
that of the Sugar Maple.

The Norway Maple is a valuable forest tree, although it is of
rather slow growth while young, but it is worthy of the attention of tree planters in our Northern States. The trees produce
seeds freely, even when of only moderate size, and can be obtained in almost unlimited quantities from trees growing in this
country. There are quite a number of varieties of the Norway
Maple, among which the following are probably the most distinct : Cut-leaved (*dissectum*), leaves regularly and deeply
divided into almost three equal parts, and of a clear, glossy
green color. Eagle's Claw, leaves cut, pointed and curled at
the point into a resemblence of an eagle's claw, hence the
name. Curled-leaved, leaves more curled, but deeply cut like
the Eagle's Claw, but still distinct. Schwerdler's Norway
Maple, leaves while young variegated with deep, reddish purple,
and sometimes the second growth in summer is similarly
marked, a handsome variety. Reitenback's Norway Maple, a
new variety somewhat like the above, but may prove to be distinct. Lorberg's Maple, leaves deeply cut but of a bright,
reddish color while young.

A. campestre.—English Field Maple.—Although this species is
very widely known as the English Maple, it is not confined to
Great Britan, but is found well distributed over Western Europe.
It is but a small tree when full grown, seldom exceeding thirty

feet in hight; consequently of no great value except for ornamental purposes. It forms a pretty little tree with roundish-
lobed leaves, twigs and smaller branches covered with corky
bark. It is well adapted to grounds of limited extent, and for
planting near buildings, as its roots do not spread to a great
distance. There are several varieties in cultivation, but none
possessing any special merit, although they may be introduced
to increase the number of varieties whenever this is an object.
There is one very pretty variety with variegated leaves, and
several others with foliage varying somewhat from the species.
About a dozen varieties are enumerated in European nurserymen's catalogues.

A. Tartaricum.—Tartarian Maple.—A small tree growing about
twenty feet high, native of Tartary. Leaves small, irregular
rounded, light colored, bark very smooth. A handsome, little,
round-headed tree. A variety of this, called the Ginnala Maple
(*A. T. ginnala*) has smaller leaves than the species, otherwise
very similar.

A. monspessulanum.—Montpelier Maple.—A small species, or
perhaps only a variety from France. It is merely a large shrub
with small palmate leaves. There are several other shrubby
maples in cultivation from Central and Southern Europe, that
are by some authors classed as species, by others, only as varieties. Among these I may mention Lobel's Maple (*A. Lobelii*),
or the Italian Maple, leaves of a pea-green color with rather obtuse lobes. This is considered by the best European authorities
as a variety of the Norway Maple. The Three-lobed Maple
(*A. trilobatum*) is another species or variety from Southern
Europe.

JAPAN MAPLES.

These Maples are of comparatively recent introduction, but
they have been with us long enough to allow of an opportunity
to test their merits, and their adaptation to the soil and climate
of this country. In these maples we have an excellent illustration of the skill of the Japanese, not only in the production,
but in the preservation and propagating of varieties of trees
and other plants indigenous to their country.

While it is not supposed that any of the Japan Maples possess
any great economic value, they are unsurpassed for ornamental purposes. In fact, their introduction has been an agreeable
surprise to the arboriculturists of both Europe and America, for

5

they are distinct from all other species and varieties of the maple. Just how many different species of maple are indige- nous to Japan is not positively known, some botanists making more and others less. There are probably four or five, and of these the Japanese have many varieties in cultivation, and some twenty or more have been introduced and pretty well tested in this country, and have, upon the whole, proved to be hardy and moderately vigorous growers for small trees or shrubs that are never expected to reach more than a few feet in hight. The varieties are grafted upon seedlings of the wild species from the forests of Japan, as none of our native maples seem to answer as stocks.

The five best recognized species of Japan Maples now in cul- tivation in this country are : *A. carpinifolium, A. Japonicum, A. Polymorphum, A. rufinerve*, and *A. epimedifolium*.

There is also another, the Colchicum-leaved (*A. colchicum rubrum*), sometimes classed as a species, but this as well as several others described in nurserymen's catalogues are not assigned to their proper places as species. They are all pretty little trees, with leaves of various forms and . colors, but the Polymorphum furnishes the greatest and most unique varieties of all. They have leaves of various shades of color, from pure green to the richest rose and crimson, and the foliage of some are so finely cut that it appears more like the feathers of some gaudy-colored bird than that of leaves of a hardy tree or shrub. Some of the varieties have leaves handsomely variegated with white, green, and yellow, and these colors are retained nearly the entire season. Words, however skilfully applied in a de- scription of these pretty little trees, would scarcely convey a cor- rect idea of their peculiar beauty, for they must be seen to be fully appreciated. *Acer rufinerve* is a curious species, with leaves resembling those of the grape, but streaked with white.

ÆSCULUS.—*Horse-Chestnuts.*

The Horse-Chesnuts have little to recommend them. except for ornamental purposes, as their wood is of a poor quality, although it is employed to a limited extent for making certain household utensils. They produce large, chestnut-like seeds, enclosed in leathery pods, which at maturity split open into three valves or divisions. There are from one to three nuts in each pod, varying in number with the different species. All the different species and varieties are ornamental, and worthy of cultivation for this purpose. I will remark here that some botanists place all the species of horse-chestnuts with smooth fruits under the generic name of *Pavia*, and the rough under *Æsculus;* but as some have fruit intermediate between the two, I have followed the most common arrangement, placing all under one generic name. The following are native species :

Æsculus Californica.—California Horse-Chestnut.—Leaves composed of five slender-stalked leaflets. Flowers white, or tinged with rose, borne in long, raceme-like panicles. Fruit large, with a few rough points on the pod, enclosing the smooth nuts. A small tree or small shrub, varying greatly in size, according to locality and soil. Wood soft, and of no value. Indigenous to California.

Æ. parviflora.—Dwarf Buckeye.—Leaves composed of from five to seven leaflets ; soft, downy underneath. Flowers white, in a long, erect raceme, appearing late in spring, or in the North about mid-summer. Fruit smooth. Seeds small. Native of the Southern States, but extensively cultivated in the Northern States as an ornamental shrub.

Æ. glabra.—Fetid, or Ohio Buckeye.—Leaflets five ; quite smooth. Flowers yellow, or yellowish white, in rather short panicles. Fruit prickly and rough. Only a moderate-sized, tall, slender tree, common west of the Alleghanies, Virginia, Tennessee, Ohio, and Missouri. Wood rather soft and of but little value.

Æ. flava.—Yellow, or Sweet Buckeye.—Leaves with five to seven smooth leaflets. Flowers yellow, in a short, compact raceme. Fruit large, smooth, or with a rough, leathery surface, the pods often assuming a bright-yellow color when mature in the fall. Native of Indiana, and southward along the Alleghany Mountains to Northern Alabama and Georgia, and westward to the Indian Territory. This is quite a variable species ; sometimes only a large shrub, while in favorable soils it grows to a large tree sixty to seventy feet high, with stem two or more feet in diameter. When planted singly, and when the branches are not crowded, it forms a globular head of handsome proportions. Wood light, soft, and not inclined to split, and used for troughs, bread trays, wooden bowls, shuttles, where a light, rather tough wood will answer. There is a native variety of this species, known as the Purple Buckeye, that has both calyx and petals tinged with purple,

Æ. Pavia.—Red Buckeye.—Very similar to the last, and by some considered only a variety, but by others as a distinct species. It is a shrub, or at best only a small tree, with bright-red flowers. A very showy and handsome plant. Natural varieties of all the above-named species occur in the forests where these trees abound, and quite a number have been secured and are now propagated for sale by our nurserymen. In addition to these natural varieties, others are constantly occurring among seedlings raised under artificial conditions. For Spanish Buckeye, see *Unganadia*.

FOREIGN SPECIES AND VARIETIES.

Æ. Hippocastanum.—European Horse-Chestnut.—This tree is supposed to have been brought from Asia, although its native country is not positively known. It has been cultivated for many centuries in Southern Europe, and for more than three hundred years in Great Britain, and is everywhere much admired as an ornamental tree. Each leaf is composed of seven leaflets, and these are of the purest green color, but not glossy or shining. The flowers are large, white, spotted with purple, produced in large, compact spikes, making a splendid appearance among the rich, green leaves. A grand ornamental tree, hardy in nearly all of our Northern States, and thriving in a great variety of soils, but succeeds best in a rather compact loam or clay. In light, sandy soils it often fails for want of moisture at the root. The "Double White Flowering" is a superb variety, bearing long panicles of very double flowers. The trees commence blooming when quite young, and seldom fail to produce flowers in great abundance. The "Cut-leaved Horse-Chestnut" is another variety with deeply-cut foliage. "Memminger's Horse-Chestnut" has its foliage sprinkled and spotted with white. In another variety the leaves are spotted with green. There are about a dozen additional varieties mentioned in the catalogues of European nurserymen, but those named above are the best.

Æ. rubicunda.—Red-Flowering Horse-Chestnut.—The origin of this tree is unknown, but it is supposed to be a hybrid between the White Flowering and some species of the Red Buckeye. Leaves of five to seven leaflets. Flowers of a bright, rosy-red color, in large panicles. One of the handsomest and most showy trees in cultivation. This tree grows to a hight of thirty feet, or more, with a close, compact form. There are several varieties, varying in habits of growth, color of the flower, or form of

foliage ; but none are superior as an ornamental tree to the
original or parent stock.

AILANTUS, OR AILANTO.

"Tree of Heaven" is a free translation of the Chinese name
Ailanto, but *Ailantus glandulosa* is the generally recognized
scientific name of a large tree of the Quassia family, native of
China, and introduced into English gardens in the middle of the
last century, and since distributed over Europe and the greater
part of America. It is a large, spreading tree, with coarse,
blunt, stiff branches, clothed in summer with long, unequally
pinnate leaves—not unlike in form those of our common Staghorn Sumach. The stem is usually very straight ; bark smooth,
of a light, grayish color. This tree was introduced into the
United States early in the present century, and attracted considerable attention as an ornamental tree. Owing to its remarkably rapid growth, its somewhat unique appearance, and
the rapidity with which it could be propagated, nurserymen
were encouraged to extol it very highly and urge it upon their
customers, far and wide. For a number of years it was in great
demand, and the "Tree of Heaven" became very popular as a
street tree in all of the larger cities and villages, besides being
extensively planted in public and private parks and gardens.
But so soon as the trees reached a bearing age, it was discovered
that the flowers emitted a most sickening and disagreeable
odor, and this called forth as loud and widespread denuncia-
tions, as had formerly been bestowed in high praise of this tree.
Thousands were cut down, but where the roots were not dug
up entire, the pieces left in the ground sprouted, and in many
instances produced a forest of trees, where previously there had
been but one. This sprouting appears to be a natural characteristic of the tree, and when the roots are disturbed, broken, or
otherwise injured in working the soil, the habit is intensified
many fold. From whence came the disagreeable odor, or from
which sex of the flowers, has been a subject that has provoked
much discussion ; but it is usually credited to the staminate
flowers borne on trees distinct and separate from those producing pistillate, and this has led some nurserymen to seek this
sex from which to propagate a stock of odorless plants. But
while this is a step in the right direction, it is not likely to be
successful, except in the hands of very close and accurate observers ; for, in fact, there are three kinds of Ailantus flowers,

instead of two, as usually claimed. The flowers of this tree are, to use a scientific-term, "polygamous," *i. e.*, having some perfect and some imperfect on the same, or on different individual trees. They are small, of a greenish color, produced in terminal, much-branched panicles, with five short sepals and five petals, and ten stamens in the sterile flower, and either none, or few, in the fertile. These three varieties of flowers may be found on different or separate trees. Those having stamens and pistils and those with stamens only, are highly odorous. The first produces seed ; the second are barren. The third kind of flowers produce pistils only, and are inodorous, but, like the first, are succeeded by fruit. From the above it may be seen that we have two odorous varieties of the Ailantus, one barren, and the other productive. But the third variety, while it produces fruit when growing in the neighborhood of either of the other two, is entirely inodorous, consequently is the only one to be propagated when the odor of the Ailantus' flowers are an objection. As the Ailantus is readily propagated by cuttings of the roots, made in the fall, and packed away in moss or clean sand during the winter, it will not be at all difficult to raise any number of inodorous trees. The pieces of roots should be kept moist and in a temperature where they will not freeze, but not warm enough to excite growth. Placing in boxes, and intermixed with sand and then buried in some dry place in the field or garden, is usually a safe way to preserve them until wanted for planting in spring. Only the smaller roots, or those of a half inch to an inch in diameter need be used for cuttings, and these may be taken from the extremity of large trees of the right sort, without destroying the parent stock.. The severed roots will produce new ones from their ends the following season, and these may be again removed, if required. Thus one tree may furnish cuttings for many years in succession, only care should be exercised in not drawing so strongly on the parent stock as to kill it.

In raising trees for a large forest, it would probably be better to resort to seedlings, instead of cuttings. The seeds grow freely when sown in the fall, or they may be kept over until spring, by storing in some moderately cool place. The Ailantus will thrive in poor light soils, where many other trees would fail, as the roots penetrate the earth very deeply, and spread a great distance, in search of nourishment. The wood is fine-grained, yellowish-white, excellent for cabinet-work and inside finish, and it also makes excellent fuel. In our more Northern States, say above latitude forty degrees, the young trees are often killed back in winter, owing to their vigorous and succulent growth. The leaves of the Ailantus furnish food for the *Bombyx Cynthia*, a species of silkworm. In Japan a cloth is made from the silk produced by worms fed on the leaves of this tree, which is not so fine in texture as that made by the common silkworm, but is much more durable. A few attempts have been made to introduce this culture in this country. No doubt it could be made successful, but at the present price of labor its profit wouldbe problematical.

As an ornamental tree, the Ailantus is certainly worthy of a place in a collection of trees ; but I do not think it worthy of much attention for other purposes, because we have many superior native species that do not possess the objectionable properties of the Ailantus. When that tree once becomes established, it is very difficult to dislodge in any other way than to clear the land, and then cultivate it almost constantly for several years in succession.

It has been urged in favor of this tree, that it will grow in the most barren soils, and where few other trees will thrive, and while in a measure this may be true, I am inclined to think that we can not only get along without the Ailantus, but it has been more of a nuisance than an acquisition to our list of valuable deciduous trees.

ALNUS.—*Alder*.

The Alders, natives of North America, are principally shrubs, or trees of moderate size, although of some species, specimens reaching a hight of seventy or eighty feet are occasionally met with in favorable locations. The flowers are very minute, monœcious, produced in catkins, the fertile ones oval, and composed of thick, woody persistent scales, enclosing small, nut-like seeds, either winged or wingless. The Alders thrive best in damp soils along the borders of streams and ponds, and some of them are valuable for planting in such situations. The timber is almost inperishable in water, and when large enough, may be employed for all kinds of cabinet work, it is largly employed for making charcoal used in the manufacture of gunpowder. The bark is employed in dyeing and tanning.

Alnus incana.—Speckled Alder, Hoary Alder, Black Alder.— Leaves broadly-ovate, rounded at the base, serrate and sometimes coarsely toothed, white and downy beneath. Generally a low shrub, but occasionally a small tree twenty or thirty feet high. Native of Northern Europe, Newfoundland, New England, and westward nearly across the continent. Wood very hard and heavy, and makes excellent fuel and charcoal, but does not grow large enough to be worthy of much attention. A variety of this species (*A. incana* var. *virescens*) is more or less abundant among the mountains of Oregon and southward to New Mexico.

A. viridis.—Green or Mountain Alder.—This, like the other older and long-known species has many synonyms in botanical works. Leaves roundish oval or oval, somewhat viscid or sticky. Seeds with a broad wing. A small shrub, native of Europe and North America, found very far to the north on this continent and southward along the mountains to North Carolina.

A. serrulata.—Smooth Alder.—Leaves obovate-acute at the base, sharply serrate with very fine teeth, smooth and green on both sides. A shrub or small tree twenty feet high. Seeds ovate and wingless. Common from New England southward to Florida. This species is also known as (*A. glutinosa*) in some botanical works and catalogues.

A. Maritima.—Sea-side Alder.—Closely allied to the above, if not identical, but some authors have claimed that it is really a distinct species, although leaves and fruit are as in *A. serrulata*. Common on the Eastern Shore of Maryland, and in Delaware. A variety of this, known as (*A. maritima* var. *arguta*), is a native of Japan.

A. oblongifolia.—Oblong-leaved Alder.—Leaves thick, oblong-lanceolate, smooth above and slightly pubescent beneath, two to four inches long. Seeds broadly-ovate, wings very narrow. A tree thirty to forty feet high, and in some instances sixty to eighty feet, with a stem two feet in diameter. Wood excellent and hard, taking a good polish. New Mexico, west to Santa Barbara, Cal. One of the largest species of Alders known.

A. rhombifolia.—White Alder of California and Oregon.— Leaves smaller than the last, or from two to three inches long, rounded or pointed at the summit, and wedge shape at the base, smooth above and thinly pubescent beneath. Seeds broadly ovate with thickened margin. Oregon to Southern

California. A tree from twenty to thirty feet high, but sometimes more.

A. rubra.—Red Alder.—Leaves thick, rusty pubescent beneath,
four to eight inches long, coarsely toothed. Seeds obovate,
surrounded by a narrow, membraneous wing. The branches
are rather stout and coarse, with bark of dark brown, dotted
with white. A tree thirty to forty feet high on the Pacific
Coast, from Sitka to Southern California, and common on the
hills about Oakland, in the vicinity of San Francisco. This is
the *Alnus Oregona* of Nuttall's "North American Sylva," and
also of catalogues.

Of foreign species of the Alder there are more varieties than
species in cultivation, but there are none which grow to a
larger size, or are of more value as forest trees than those found
indigenous to North America, in fact, the Alders of both conti-
nents seem to be very closely allied and probably all spring
from the same original stock. Some very handsome varieties
are cultivated in nurseries, especially those with finely cut
leaves.   These are propagated by grafting, although all the
species and varieties of the Alder may be readily propagated by
cuttings planted in low, moist soils. The seeds also germinate
readily, and may be gathered and treated the same as those of
the maple, and similar forest trees.

AMELANCHIER.—*June-Berry, Service-Berry, Shadbush.*

Of this genus we have only one indigenous species that grows
large enough to be classed among trees. The flowers are small,
pure white, produced in long racemes, and in such great
abundance in early spring that the trees become conspicuous
and attractive objects scattered along the banks of thousands
of the small streams and rivers throughout the country, for
this species, or some of its varieties inhabit almost every square
mile of forest from Hudson's Bay in the north, southward to
Florida, and westward to the Pacific, and even growing at an
altitude of ten thousand feet in the Rocky Mountains.

Amelanchier Canadensis.—Eastern Shadbush.—Leaves simple,
sharply serrate. Flowers white. Fruit small, berry-like, roundish,
purple when ripe, sweet or sprightly, sub-acid, edible. A small
tree, but sometimes fifty feet high, with stem a foot or more
in diameter, wood hard, very heavy, and resembling that of the
apple tree. This is an exceeding variable species, and it runs
into many forms or varieties, to which distinct names have

been given in botanical works, and nurserymen's catalogues. The best known of these are : Var. *Botryapium*, leaves ovate-ob-long, sometimes heart-shaped, (fig. 26). Flowers larger than the above and more showy. Var. *oblongifolia*, leaves oblong, while downy when young, racemes and petioles shorter than those of the last. Var. *rotundifolia*, leaves broader and more oval, sometimes nearly round, and the racemes of flowers short. Var. *oligocarpa* (var. *pumila*) of catalogues, leaves smooth,

Fig. 26.—DWARF JUNE-BERRY.

narrow oblong, racemes of only three or four flowers. A very dwarf shrub, seldom more than three or four feet high. Fruit quite large and usually more abundant than on the taller grow-ing varieties.

A. alnifolia.—Alder-leaved Shadbush.—Leaves broadly-ovate or rounded, obtuse at both ends, or somewhat cordate at base. Racemes of flowers short. A low shrub, perhaps only a variety of *A. Canadensis*, found west of the Rocky Mountains, and northward to British Columbia.

AMYRIS.—*Torch Wood.*

Trees and shrubs of Tropical America, with opposite com-pound leaves, mostly of a single pair, or trifoliate. pinnate.

Only one species reaching as far north as the United States, and this only in Southern Flordia.

Amyris sylvatica.—(*A. Floridana*, Nutt.) Florida Torch-wood.—Leaves small with divided ovate pinnæ. Flowers with four white petals. Fruit purple, containing one seed or nut. Wood yellowish-white, close-grained and susceptible of a high polish, and the wood is also fragrant, having a balsamic odor. A small tree of no value except for cultivation in tropical climates.

Andromeda-arborea.—See Oxydendron.

ARALIA.—*Angelica Tree.*

There are several indigenous species of plants belonging to this genus, but only one with a woody stem, the others are herbaceous plants.

Aralia spinosa.—Hercules' Club.—Leaves very large, crowded at the summit of the stem, bipinnatedly compound. Flowers minute, white, in very large panicles, succeeded by small, berry-like, black fruit. The stem and branches are very prickly, especially while young. A well-known shrub or small tree, often cultivated in gardens on account of the tropical appearance of its immense compound leaves. Not quite hardy in the more Northern States, the stems are often killed down in winter, but the roots usually survive, and throw up vigorous shoots in the spring. Native of Southern Pennsylvania, Kentucky, and southward to Florida, and westward to Texas. In Southern swamps it sometimes reaches a hight of fifty feet, with a stem a foot in diameter. A tree desirable only as a curiosity or for ornament. The roots if disturbed throw up suckers in great numbers. Readily propagated from seeds or cuttings of the roots.

· There are several Asiatic species and varieties, several of which are now quite common in gardens. The *Aralia chinensis*, also known as *A. canescens, A. elata*, also *Dimorphanthus elatus*, Miguel, or *D. manschuricus*, Maximowicz, is as hardy as our indigenous species, and the flowers are in larger panicles. A Japanese species, *Aralia Japonica* of Thunberg, and *Fatsia Japonica* of Decaisne and Planchon, has yielded several handsome varieties with variegated foilage, but these are of more interest as ornamental shrubs, than as useful trees.

ARBUTUS TREE.—*Madrono.*

A genus of trees or shurbs containing but five species, principally belonging to the temperate regions of the Old World, the most familiar of these is the Strawberry-tree (*A. Unedo*) of which there are several varieties. There are also two or three species found in Mexico, and one or two Asiatic species, but the one of the most interest to the arboriculturists is the Madrono, found on the west coast, or

Arbutus Menzelesii.—Menzies' Arbutus.—Its synonyms are *A. laurifola*, Lindley. *A. procera*, Douglass. *A. Texana*, Buckley. Leaves oval or oblong, either entire or serrulate, pale beneath, bright green above. Flowers white, in dense racemes. Fruit a berry, dry, orange colored with a rough surface, not edible. A splendid, large tree, eighty to one hundred feet high, with a stem two to three feet in diameter in Northern California, but smaller southward. Wood white, very hard, but brittle. A tree is mentioned in Geological Survey of California, Botany, Vol. I., found in Marin County, measured twenty-three feet in circumference at the smallest part of the stem below the branches, and some of the branches were three feet in diameter. South of San Francisco Bay it is usually a small, spreading tree or shrub. From Puget Sound southward to Arizona, and eastward to Texas, As this tree appears to thrive best in cool climates—at least it grows larger in Northern California than anywhere south—it may prove of value as an ornamental tree in our Atlantic States.

ARCTOSTAPHYLOS.—*Manzanita.*

Shrubs or small trees, with alternate leaves of a leathery texture, nearly entire or with fine irregular teeth. Flowers white, or rose-colored in terminal racemes, succeeded by small, plum like fruits, containing five to ten separate or separable long seed-like stones. In propagating these plants, the seed should not be permitted to get thoroughly dry. The fruit may be placed in heaps or in masses, until the pulp becomes softened, then the seed washed out and either sown immediately, or put away in moist earth or sand, until the time arrives for sowing in spring or fall.

The following seven species are only shrubs : *A. Andersonii*, six to ten feet high. Fruit reddish, Santa Cruz, Cal. *A. tomentosa*, two to six feet. Fruit red, smooth. Used for making a cooling sub-acid drink. From Puget Sound to Southern Califor-

nia on dry hills. *A. nummularia.* Erect, but only one or two
feet high. Very leafy, like the Dwarf-box. *A. Uva-Ursi* (Bear-
berry). Trailing leaves, thick and evergreen. This is the *Kinnikinick* of the Western Indians, and is found on rocky, bare
hills throughout the northern part of Europe, Asia, and
America. *A. pumila*, is a closely allied species to the last, but
stems erect. California. *A. Alpina*, dwarf, tufted. Fruit
black. Alpine region of Europe, and North America. *A. polifolia*, erect, five to eight feet high. Fruit dark purple, minutely
warty. Southern California.

Arctostaphylos bicolor, Gray.—An erect shrub, three to four feet
high. Flowers rose-color. Fruit small, the size of a pea, yellow,
turning to red, and from one to five seeds in each. California'
San Diego, and near Monterey.

A. pungens.—California Manzanita.—Leaves with a long stem,
oblong-lanceolate or oval. Flowers crowded in a short raceme.
Fruit reddish. A small tree, twenty to thirty feet high, but on
the mountains only a small shrub. Wood very hard, heavy,
and the color of mahogany. Excellent for the finer kinds of
cabinet work. Southern Utah, Arizona, California, and Mexico.

A. glauca.—Leaves very stiff, oblong, slightly heart-shaped.
Fruit red, large, smooth, nut enclosed in a thin pulp. This is
known in California as the Great-berried Manzanita, as the fruit
is sometimes three-fourths of an inch in diameter. A small
tree twenty feet high, with stem sometimes a foot in diameter.

ARDISIA.

A genus containing many species of handsome evergreen
shrubs, or small trees, native of tropical countries, valued for
their handsome foliage, small but showy flowers, and pretty
berries, which are usually very persistent, remaining a long
time attached to the plant. One species in the United States.

Ardisia Pickeringia.—Leaves smooth, oblong-ovate, obtuse, entire two inches long, narrowed at base, into a short petiole, pale
beneath. Flowers small, in short terminal racemes. A large
shrub or small tree, twenty to thirty feet high. Southern
Florida, west to Mexico, also in the West Indies. *Ardisia Japonica* is quite a favorite for green-house culture, on account of
its bright and persistent berries. All the species easily multiplied from seed or cuttings of the young shoots.

ASIMINA.—*Papaw, Custard Apple.*

Small trees or shrubs, with deciduous leaves. Fruit large, in clusters, pulpy, containing several large flattish seeds. The Papaw is an edible fruit, and those who become accustomed to its use consider it excellent and well worth cultivating. The species are:

Asimina triloba.—Leaves oblong-ovate, pointed, covered with a rusty pubescence, and the young branches are slightly covered with the same, but become smooth with age. The leaves are quite large, sometimes nearly a foot long, and half as wide on young vigorous specimens. Flowers are of a peculiar form, as shown in fig. 27. The outer petals round ovate, greenish-yellow at first, but changing to dark purple. Fruit banana-shaped or oblong, three to four inches long, consisting of a sweetish pulp, containing several large flattish bony seeds. A very handsome small tree, sometimes thirty or more feet in hight. Wood

Fig. 27.—FLOWERS OF PAPAW.

rather light and spongy; not valuable. The fruit might be greatly improved by cultivation, and new varieties produced as with other similar native fruits. Found sparingly in Western New York, more abundant westward to Iowa and southward to Florida. Readily propagated from seed or suckers, which usually spring up more or less abundantly from the roots.

A. parviflora.—Small-flowered Papaw.—A small shrub South, in dry soils. Leaves smaller and thicker than the last, and flower only a half inch broad. Fruit small, oblong, or pear-shaped.

A. grandiflora.—Large-flowered Papaw.—Also a small shrub,

South, with leaves only two to three inches long. Flowers
with outer petals two inches long and yellowish-white. Fruit
small, often containing only one seed.

A. pygmæa.—Dwarf Papaw.—A small shrub. Georgia and
Florida. Flowers small, appearing late in the spring or summer from the axils of the leaves of the season.

The Custard Apple of the West Indies (*Anona glabra*), may
be mentioned here, as it is occasionally found in southern
Florida, where it may have been introduced by the Indians, or
escaped from some of the islands and washed ashore or seeds
dropped by birds. It is a small tree, and only of value in a
tropical climate.

AVICENNIA.

Low evergreen shrubs or trees, with long creeping roots,
forming dense and almost impenetrable thickets in saline
marshes along the sea-shore in tropical or semi-tropical climates.
Two species are found in Florida and along the Gulf to westward. Only one of these, "The White Mangrove" (*A. nitida*,
Jacq., *H. oblongifolia*, Nutt.), reaches a hight of twenty feet, and
this one very seldom; consequently the genus is of no especial
interest except to the botanist or residents of tropical countries.

BETULA.—*Birch*.

A widely distributed genus, containing many large-growing,
useful and ornamental species of trees, the bark and wood of
some highly aromatic. The twigs and younger branches are
generally rather slender and very flexible, giving to the trees a
very graceful habit, a characteristic of the entire genus,
whether trees or shrubs. They thrive in a great variety of
soil, but succeed best in one that is moist. The flowers are
monœcious, that is, the sexes are produced separately, pistils in
one and stamens in another, but both in scaly catkins on the
same tree. Seeds small, nut-like, surrounded by a wing. They
are propagated by seeds, which ripen in autumn, budding and
grafting, and in the dwarf species by layers. Our indigenous
species are as follows :

Betula alba.—Var. *populifolia*.—White Birch, Gray Birch.—
Our native White Birch is now considered by botanists as only
a variety of the European *B. alba*, hence the use of two botanical names as above. Leaves small, somewhat triangular and
tapering, very smooth and glossy. Stem with chalky white

paper-like bark, readily peeling horizontally in thin sheets. Wood very white, firm, close-grained, easily polished; extensively used in the manufacture of spools, shoe-pegs, and other similar purposes. In Russia the oil from White Birch is said to be used to give to Russia leather the peculiar aromatic and lasting qualities, and when dissolved in alcohol is said to be excellent for preserving and water-proofing various fabrics. A small, rather slender tree thirty or more feet high, growing in poor, sandy, and gravelly soil, also in cold, moist soils near ponds, swamps, and along the banks of streams. Common almost anywhere in the Northern States and Canadas, and also along the mountains southward.

B. papyracea.—Paper or Canoe Birch.—Closely allied to the White Birch, but a much larger tree. Leaves ovate or heart-shaped, dark-green on the upper side. The bark papery and readily separated into large sheets impervious to water, hence its extensive use by the Indians for making tents, baskets, canoes, and various domestic utensils. Wood white, compact, hard, making excellent fuel, and is also used for the same purposes as the White Birch. Extensively exported from the New England States and Canada. Common throughout British America, the Northern States, and westward to Dakota.

B. lenta.—Black Birch, Sweet Birch, Mahogany Birch, Cherry Birch.—Leaves oblong-ovate and somewhat heart-shaped, finely and doubly serrate. Bark dark-brown, close, not peeling readily ; very aromatic. Wood of a reddish color, fine grained, compact, excellent for cabinet work and fuel. A large tree fifty to sixty feet high, with stem two feet in diameter. Throughout the Northern States and Canadas, in moist soils, and southward to Georgia in the mountains. A valuable tree for planting in moist soils in cold climates.

B. lutea.—Yellow Birch, Gray Birch.—Leaves of a dull green color, oblong-ovate, rarely heart-shaped. Bark less aromatic, and of a grayish color, separating in very thin layers. Wood similar to that of the Black Birch, but can be obtained of a larger size, for the Yellow Birch is said to be the largest deciduous tree found north of the Great Lakes, growing seventy to eighty feet high, with a stem three to four feet in diameter. From Newfoundland to Dakota, Manitoba, and southward in the mountains of North Carolina. A valuable forest tree, and worthy of extensive cultivation in the Northern States.

B. nigra.—Black Birch, River Birch, Red Birch.—Leaves rhombic-ovate, whitish beneath, and the small twigs of a rusty color. A small slender tree along the banks of streams, from New England southward to Florida, and westward to Texas. More abundant South than in the North.

B. occidentalis.—Western Birch.—Leaves thin, broadly-ovate, acute, abrupt, or somewhat rounded at the base, one to one-and-a-half inches long. Wings of seed very broad. Described by Watson in Botany of California as a small tree, ten to twenty feet high, in the eastern canyons of the Sierra Nevada at an altitude of from four thousand five hundred to ten thousand feet. Extensively employed for fuel and fencing. Found in Washington Territory to the Saskatchewan, and southward in the Rocky Mountains to New Mexico. *B. glandulosa*, is a low shrub, inhabiting the same region as the last and farther north.

There are also several cultivated varieties of our native species of Birch, the best known of which are the Cut-leaved (*laciniatum*), and the Weeping (*pendula*), these are propagated by grafting or budding on stocks of the more common kinds. Of foreign species there are quite a large number, but there are none among them in any way superior to our native species as forest trees.

BOURRERIA HAVANENSIS, Miers.

A small tree found on the Florida Keys and in the West Indies. It is one of those unfortunate plants that has more names than merits. It is the *Ehretia Havanensis* of Willdenow, and is described in Chapman's Flora of the Southern States under the name of *Ehretia Bourreria*, p. .329. This species may be found in botanical works under some seven or eight different names, and a variety (var. *radula*), has five. It is of no special interest further than adding one to the number of trees and shrubs indigenous to the United States.

BUMELIA, Swartz.—*Ironwood, Buckthorn.*

Spiny shrubs or small trees with very hard wood. Leaves deciduous. Flowers small, white or greenish-white in the axils of the leaves. Fruit an ovoid one-seeded berry, and edible.

Bumelia tenax, Willd.—Leaves broadly-lanceolate or spatulate, one to three inches long. Flowers in clusters. A small tree, twenty to thirty feet high with divergent branches. North Carolina to Florida, in sandy soils.

B. lanuginosa, Pers. — Leaves oblong-obovate. Flowers in clusters of six to eighteen. A tree sometimes forty feet high, and not so spiny as some of the species. Missouri and southward to Texas and eastward to Florida. A variety of this (*B. macrocarpa*, Nutt.), has leaves less than an inch long. Fruit edible, and quite large.

B. lyeloides, Gaertn.—Leaves quite smooth, obovate-oblong, two to five inches long, often whitish underneath when young. A low shrub, but sometimes a tree twenty to thirty feet high. Illinois to Texas, and eastward to Florida.

B. cuneata, Swartz.—Leaves quite variable in form; very long lanceolate or broadly obovate; an inch to an inch-and-a-half long, very thick and fleshy. A small tree twenty to thirty feet high, but more commonly only a low shrub. Florida, West Indies, Texas, and Mexico.

BURSERA— *West India Birch.*

Tropical American trees, yielding a transparent green rosin, readily dissolved in alcohol and occasionally used as a varnish. Only one species found in the United States and this is the

Bursea gummifera, Jacquin.—Leaves unequally pinnate, three to five leaflets. Flowers small in axillary racemes. Fruit a drupe the size of a small hazelnut. Seed a small white nut, each containing one kernel. The Spanish name is *Almicigo* or Mastic Tree. A large tree in Southern Florida and in the West Indies. Wood soft and brittle.

CALYPTRANTHES, Swartz.

A genus of small evergreen trees indigenous to the West Indies and Brazil. Flowers very minute but numerous, usually in axillary or terminal branching racemes. We have one species:

Calyptranthes Chrytraealla. — Forked Calyptranthes. — Leaves ovate or ovate-lanceolate, rather blunt-pointed, smooth above but pubescent beneath. Flowers whitish, minute. Berry dry, round; one or two-seeded. Wood very hard, and in Jamaica considered an excellent timber, but the tree does not grow to a large size, and the stem is seldom more than a foot in diameter. Found at Key West, Florida, and in the West Indies.

CARPINUS.— *Blue Beech, Water Beech.*

Tall shrub or small tree, widely distributed in North America, only one indigenous species.

Carpinus Americana of Michx. ; or C. caroliniana of Walt.— Leaves ovate-oblong, pointed, doubly serrate, very smooth and thin, resembling those of the Common Beech (*Fagus*). Sterile flowers in rather dense catkins, and fertile ones in little slender, loose catkins, with a pair of three-lobed bractlets, one on each side of the small nut-like seed, which ripens late in the autumn. Shrubs and trees, from twenty to forty feet high, often a number of stems springing from the same root. Common in swamps, and along the banks of streams from Nova Scotia, westward through the Canadas and Northern States, and southward along the Alleghanies to Georgia, and in the rich woods of Florida. Bark smooth and of a grayish color, stem often deeply furrowed. Wood very white, hard, close-grained and exceedingly tough. Extensively used by the early settlers of our Northern States for making brooms, as the wood is so tough that it is easily divided into very thin and narrow strips for the brush of the home-made broom. A blue-beech withe will last almost as long as iron wire, and an ox-gad made of a blue-beech sprout is nearly equal to a leather one. There may be many of my readers who have seen an armful of the same kind of implements of torture, brought into the country school-house and placed near the fire or drawn through the hot ashes on the hearth, to take the frost out and increase their flexibility and toughness of the rods, which were once considered very important aids in preserving the discipline of a district school. There is one European and an oriental species of *Carpinus*, but neither are of any special value as timber trees. For another tree closely allied to the *Carpinus* botanically, but otherwise very distinct, see *Ostrya Virginica*.

CARYA.—*Hickory.*

The hickories are a very important genus of North American trees, supplying almost every branch of mechanics with very tough timber, and for fuel it has no superior. They are principally trees of large size, with alternate, odd-pinnate leaves, which usually assume a golden hue in autumn. The flowers are monœcious, the fertile ones very minute, opening at the apex of the embryo nut, and the sterile or male ones in long, pendulous catkins. Seed, a nut enclosed in a thick or thin, four-valved epicarp or husk. All are readily propagated from the nuts, which should be stripped from their outer husk soon after they fall, and then buried in heaps, mixed with sand or soil, and left exposed to frosts during the winter, to be sown in drills in spring, or they may be planted in rows at the time of gathering. When planted in seed-beds composed of rather light or sandy soil, the seedlings will produce a greater number of small, lateral roots than when the nuts are planted in heavy clay. The seedlings may be transplanted when one or two years old, and a portion of the tap root removed as directed in a previous chapter. When treated in this way, all the hickories are as readily and safely transplanted, as the chestnut and similar forest trees. Propagation by budding and grafting has not been very extensively or successfully practiced by our nurserymen. By securing good, thrifty seedling stocks, and then grow them in pots for a year, or until they are well

Fig. 28—THICK SHELL-BARK   Fig. 29.—CROSS SECTION OF THICK
        HICKORY.   SHELL-BARK HICKORY.

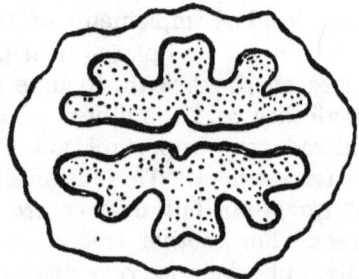

established, a fair degree of success may be obtained in grafting the hickory in propagating houses or in frames. In warm climates the propagation of nut-bearing trees of all kinds appears to be attended with far greater success than in cold ones. Loudon in referring to the subject in Vol. III., *Arboretum* and *Fruticetum*, p. 1431, says : "Much has been written on the subject by French authors, from which it appears that in the north of France, and in cold countries generally, the walnut does not bud and graft easily by any mode ; but that in the south of France, and north of Italy, it may be budded or grafted by different modes with success." The same may be said to be true in this country, and while both the hickories and the walnuts are not readily propagated by budding or grafting in the nursery at the North, they are in the South, as many correspondents have assured me. Varieties may, however, be multiplied, by exposing a portion of the roots of the large trees to the air and light, and from the exposed parts sprouts will appear, and when these are two or three feet high may be taken up and transplanted with a section of the parent root attached.

Carya alba, Nutt.—Shell-bark or Shag-bark Hickory.—Leaflets five to seven, usually five, lanceolate oblong, the upper three much the largest. Fruit flat or depressed at top, nut white, roundish, or slightly four-angled, with a sharp point at the apex. Thin shelled, and kernel sweet and excellent. Nuts highly prized, and always in demand. Wood heavy, tough, and elastic, highly valued by the manufacturers of agricultural implements, carriages, etc., etc. A large tree, often eighty feet high, and stem two to three feet in diameter. Bark shaggy or scaly. This is not only a noble and valuable forest tree, but a superb ornamental tree, which deserves far more attention than has ever been given it. More or less abundant in all of our Northern States and the Canadas, and westward to Nebraska. Also occasionally found as far south as the northern part of Georgia.

C. sulcata, Nutt.—Western Shell-bark and Thick Shell-bark Hickory.—Leaflets seven to nine, obovate-oblong, slightly downy beneath. Fruit very large, oval, somewhat four-angled above. Nut oblong, dull white or yellowish, with a point on both ends, as shown in figure 28, which is of the exact and an average size of some nuts of this species I received from Ohio. The shell is also very thick, as shown in a cross section of the same nut, figure 29. The kernel, although small in proportion to the size of the nut, is sweet-tasted and good. A large tree, with a rough bark somewhat scaly. Wood heavy, tough, and excellent, but the heart-wood is more like that of the next species than that of the last. A more common tree west of the Alleghanies than east of them, but was formerly quite abundant in Western New York and southward through Pennsylvania to North Carolina.

C. tomentosa, Nutt.—Mocker-nut, White-heart Hickory.—Leaflets seven to nine, mostly seven, large, oblong-ovate, sharp pointed, lower surface downy when young. Fruit large, round, usually with very thick, hard husk. An exceedingly variable species.

Kernels sometimes sweet and good, then again scarcely eatable. The "King nut," known in the Genesee Valley, N. Y., is said to belong to this species, although its shell is quite thin, and the kernel large and excellent. The wood of this species is as variable in quality as the nuts are, and while, as a rule, it is very white, heavy, and only moderately tough. I have cut trees that gave the straighest grained, and toughest hickory wood I ever handled. A very tall but slender tree, with a rough, deeply-furrowed bark on old trees, but does not split off in strips, as in the last two species. More common on high, dry ridges, than in low lands, plentiful in the sandstone regions of New Jersey and southward to Florida. Also in New England, Canada, and westward.

C. olivæformis.—Pecan-nut.—Leaflets thirteen to fifteen, oblong-lanceolate, taper-pointed. Fruit cylindical oblong, nut olive-shaped, yellowish-brown, shell very thin, kernel sweet and delicious. The nuts are usually a little over an inch long, and quite regular in form ; but an occasional tree will produce much larger nuts or of the size shown in figure 30, which was made from a fair average number of these nuts received from a correspondent in Louisiana.

The Pecan-nut tree grows to a large size in the bottom lands along the rivers in the South and West. Wood similar to that of the Shell-bark Hickory, and very valuable. Southern Illinois is its northern limits in its wild state, but it has been cultivated in more northern localities with rather indifferent success.

Fig. 30.—PECAN NUT.

C. porcina, Nutt.—Pig-nut, Brown Hickory.—Leaflets five to seven, ovate-lanceolate, smooth. Fruit ovate, oblong, or pear-shaped, quite variable in size and form. Husk thin, opening at the top, often remaining on the thick shelled nut all winter. Kernel usually bitter, but sometimes pleasant flavored. A large tree, with smooth bark, common in the same regions as the White-heart Hickory, and the wood very similar.

C. amara, Nutt.—Bitter-nut, Swamp Hickory.—Leaflets seven to eleven, oblong-lanceolate, pointed, slightly downy when young. Buds on the small, slender twigs, yellowish in winter,

resembling those of the butternut. Fruit globular, husk very
thin, nut yellowish, thin shelled. Kernel intensely bitter.
Wood rather soft, white, but often quite tough. A small, slender tree of a graceful habit when allowed room for full development of its branches. Common in low, moist ground, from
Canada to Florida, and westward to Texas.

C. aquatica, Nutt.—Water Hickory.—Leaflets nine to eleven,
oblong-lanceolate, pointed, smooth. Fruit roundish, four ribbed,
husk thin, nut flattish, four angled with thin shell, and kernel
very bitter. A small tree with rough, somewhat furrowed
bark. Wood similar to the last, and I may add a closely allied
species, and perhaps only a southern variety of it. From North
Carolina south and westward.

C. myristicæformis.—Nutmeg Hickory.—Leaflets five, ovate-
lanceolate, smooth, the terminal ones sessile. Fruit oval,
rough; nut of same form, pointed, shell hard, furrowed, resembling the nutmeg, hence its name. A small tree in the
swamps and low grounds. South from South Carolina, westward to Louisiana. •*Carya microcarpa* of Nuttall, is now con-

Fig. 31.—HALES' PAPER-SHELL
HICKORY.

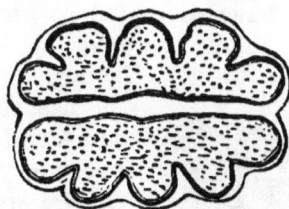

Fig. 32.—CROSS SECTION OF HALES'
PAPER-SHELL HICKORY.

sidered by some of our highest botanical authorities to be only
a variety of the Shell-bark Hickory (*Carya alba*).

There are hundreds of quite distinct varieties of the hickories
to be found in our forests, and some of them are well worthy
of preservation and propagation. But in our northern climate,
budding and grafting the hickory is seldom successful, although
several of my correspondents assure me that they find little
more difficulty in grafting hickories than they do the apple or

pear, and in the open grounds and upon stocks of large size. The difference is probably due to climate, for we certainly have as skillful propagators in our northern nurseries as there are to be found anywhere, but for some reason few have been successful in propagating the hickory by any of the well-known methods of budding or grafting, either under glass or in the open air. The cion will often unite and grow slowly for a season or two, and then die, the union appearing to be an imperfect one. One of the most distinct and valuable varieties that has ever been brought to my notice is the "Hales' Paper Shell Hickory Nut." I first became acquainted with this variety some fifteen years ago, and the next season described it under the above name. It is a variety of the common Shell-bark Hickory, and the tree is now growing with several others of the same species, near the Saddle River, on the farm of Mr. Henry Hales, about two miles east of Ridgewood, Bergen Co., N. J. The tree is a large one, and produces a fair crop of nuts annually. This handsome variety is well represented in the accompanying illustration, figure 31, showing a side view, natural size, and figure 32 a cross section, showing thickness of the kernel, and the extreme thinness of the shell. The general appearance of this nut is quite similar to some of the varieties of the English walnut, the surface being broken up into small depressions, instead of angles and corrugations as usually seen in the large varieties of the shell-barks. Of all who have attempted to propagate this fine, and I think I may say unique variety, by budding and grafting, Mr. J. R.

Trumpy, of Flushing, N. Y., is the only one who has thus far been successful, he having succeeded in making a little less than one hundred live.

If the space at my disposal would admit of it, I should be pleased to refer to several other varieties that I have obtained from various parts of the country, but as it will not, I must be satisfied with referring to one which in the

Fig. 33. DEFORM-ED HICKORY NUT.

way of a freak of nature can scarcely be excelled. This variety comes from Connecticut, where a very large tree of the Shell-bark Hickory annually produces a good crop of excellent nuts of the form and size shown in figure 33, one lobe or side of each nut fails to fill out, the nuts otherwise being perfect.

CASTANOPSIS.—*California Chestnut or Chinquapin, Golden-leaved Chestnut.*

A curious genus of trees found in Eastern Asia and adjacent islands. Botanically, the genus is intermediate between the true chestnuts and oaks, represented in this country by one species on the Pacific Coast.

Castanopsis chrysophylla, A. DC.—California Chestnut, Chinquapin.—Leaves evergreen, thick and leathery, oblong or lanceolate, two to four inches long, pointed, with short petiole, green above, and densely scurfy beneath. Fruit with stout spines, one half to an inch long; nut usually solitary, somewhat triangular, and shell firm and hard. A small tree from thirty to forty feet high, but in some situations only a small, low shrub. From Oregon to Monterey, and in the Sierra Nevada at an altitude of six thousand feet, will probably thrive in some of the Middle, and all of the Southern States.

CASTANEA.—*Chestnut.*

A well-known genus containing a limited number of species, of which there are many varieties. The staminate (male) flowers are yellow, and produced in long, pendulous catkins, and the pistillate in a bell-shaped involucre, which, as it enlarges, becomes a globose, prickly fruit, enclosing one to three ovoid, brown nuts. In figure 34 is shown a bunch of chestnut flowers, the long catkins being the staminate, and above these on a branching flower stem is shown four of the small embryo burs or fruit, the fertile flowers being situated on their apex. On this flower stem, and above the embryo fruit there are also staminate flowers only partly developed, while those below are in full bloom. If the female flowers open too late, or fail to be fertilized by the staminates in the large catkins, they are very certain to be by the later ones situated above them on the fruiting branches as shown. The species are as follows:

Castanea vesca.—European Chestnut.—Leaves oblong-lanceolale, pointed, coarsely serrate, smooth on both sides. Nuts large, two to three in each burr. The texture of the nuts are rather coarse, with very little sweetness, and to make them more palatable they are either roasted or boiled. The European Chestnut has not as yet been very extensively planted in this country, as it is not, as a rule, quite as hardy a tree as the

6

Fig. 34.—CHESTNUT FLOWERS.

native Chestnut, still by careful selections from the most hardy seedlings, and the propagation of these, we could no doubt secure very valuable varieties well adapted to a wide range of country. Several such promising varieties are now being propagated and promise to be acquisitions to our list of nut-bearing trees. One of the best and most promising of these varieties with which I am acquainted is call Numbo, and was selected from a large number of seedlings raised by Mr. Moon, of the Morrivsille Nursery, Bucks County, Pennsylvania, some thirty years ago. The original tree is now about forty feet high, and produces a large crop of nuts every year. This variety is now being disseminated, and if successful others will no doubt follow. The European varieties appear to succeed better when worked on our native stocks than on their own roots.

C. vesca (var. *Americana*).—American Sweet Chestnut. — Leaves more acute at the base, and not usually as large or as thick. Nuts smaller, more delicate, shell thinner, and kernel much finer grain, and sweeter than the European. The nuts are in great demand in the fall and early winter, but are so delicate that they soon wither up if kept in a dry place, and become mouldly if kept in a moist and warm one, but those who may wish to extend the season for these nuts, may readily do so by mixing them with clean, moist sand, which if buried in some dry place in the open ground, where they will be kept cool, and neither too dry or too wet, may be preserved in good condition a long time. If put in small boxes a supply of fresh chestnuts may be kept up from fall to late in spring. I have practised this method of preserving chestnuts for many years, and have never failed to carry them through the winter in a sound condition, and in spring they were in good order for eating or planting. It is not necessary to bury the boxes containing the nuts below the reach of frost, but merely so deep that they will not be effected by every change in the weather. The wormy and imperfect nuts will of course decay, and it is a good plan to keep the nuts for a few weeks after gathering, and then carefully select the good ones before putting away in sand. The chestnut is one of our most valuable forest trees, growing to an immense size in favorable situations. The wood is rather coarse-grained, only moderately tough, but strong and durable. It is of a light-yellowish or brown color, and is much used for fence rails, posts, stakes, railway ties, also for beams, joists, and other parts of buildings, although it is very liable to warp when seasoning, and for this reason is objectional for either. hewn or sawed timber, to be used in the frames of buildings. It is extensively employed in the manufacture of furniture, and inside finishing of dwellings and other buildings, but must be very thoroughly seasoned before used. It makes very poor fuel, not worth half as much as hickory, as it burns slow, snaps disagreeably, and throws out little heat. The wood of old trees is quite durable when used for railway ties and fence posts, but the young trees of from six to twelve inches in diameter will seldom last more than eight or ten years, unless charred or coated with coal tar, or some similar wood preservative. I have used many hundred of chestnut fence posts of from five to ten inches in diameter, and must say that this rapid decay, even in very dry soils, has somewhat surprised me, inasmuch as the chestnut is so frequently recommended for such purposes on account of its great durability. This tree, however, is well worthy of extended cultivation wherever the climate and soil will admit of it, for it is of exceedingly rapid growth and may be transplanted as successfully as almost any of our cultivated fruit or forest trees. It is said to avoid limestone regions, and stiff, clayey soils, but is at home on slaty or granite ridges, and in sandy soils, whether high or low. On the red sandstone ridges of New Jersey and southward, the Chestnut trees are abundant, and reach a large size. The nuts are quite variable in size and form, and there are occasional natural varieties that are almost equal in size to the European chestnut. These should be selected in preference to the smaller ones, by those who desire to raise trees for producing nuts.

The Chestnut is found in Southern Maine, west to Michigan, and southwest to Arkansas. Also in all of the Eastern States and south to Florida.

C. pumila.—Dwarf Chestnut, Chinquapin.—Lance-oblong leaves, downy beneath. Nuts small, round, solitary, or only one in each burr, very sweet and fine grained. A handsome little tree, with a roundish head, thirty to forty feet high, growing on sandy ridges, from Pennsylvania and Southern Ohio, southward to the Gulf of Mexico. Hardy in Northern New Jersey, and about New York City where an occasional tree planted a half century ago is seen. By grafting the Chinquapin on stocks of the common Chestnut, the growth will be greatly improved, and the trees will come into bearing much earlier than when raised from seed. Wood very similar to the common Chestnut, but the small branches and twigs are tougher.

The Chinese and Japanese Chestnuts are probably varieties of the European species (*C. vesca*), and the one recently introduced under the name of *Castanea Japonica* is a handsome little tree, coming into bearing when quite young, is very prolific, and the nuts are quite large. The tree, so far as it has been tested, appears to be as hardy as the common Chestnut. Only a few specimens have as yet fruited in this country, consequently, I can say little more of it than that it promises to be an acquisition. There are also quite a number of ornamental varieties in cultivation, one of the best known of these is the Cut-leaved (var. *laciniata*), a handsome tree with the margin of the leaves finely cut or divided.

CATALPA.—*Indian Bean Tree.*

Shrubs or trees with large, simple leaves, branches rather coarse and stiff. Flowers bell-shaped, appearing late in spring. Seed flat with fringed wings, produced in long, slender pods. All the species readily propagated by seeds, layers and cuttings of the ripe wood.

Catalpa bignonioides, Walt.—Common Catalpa.—Leaves large, heart-shaped, pointed, downy beneath. Flowers tubular, bell-shaped, somewhat five-lobed, an inch long, white, flecked on the inside with yellow and purple ; appearing late in spring in large, open, terminal panicles as shown in figure 35. Seed-pods very long, a foot or more. Seeds flat with narrow or broad fringed wings. A small or large tree, according to soil and location, with very coarse, stubby, cane-like shoots and branches, usually a round-headed tree when given room, but when growing in forests the Catalpa assumes an erect and sturdy habit, growing to a hight of sixty or more feet, with stem two to four feet in diameter. Wood light, only about one half the weight of common hickory, but close-grained and susceptible of a fine polish, very durable and valuable for fence posts, railroad ties, and all similar purposes. A rapid growing tree, especially when young. As a shade tree it has been extensively planted in our Eastern States, and on dry, well-drained soils it is quite hardy in localities where the temperature of winter does not go more than twenty degrees below zero, although I have known it to withstand twenty-seven degrees below in my neighborhood without injury, even when many of the native

Fig. 35.—CATALPA FLOWERS, (one-third natural size.)

red cedars suffered severely, and thousands were killed. Seeds and plants of the Catalpa tree of our Eastern States were sent to Europe very early (1726), and Loudon in 1838 refers to a large number of specimens growing in Great Britian and various countries of Europe, giving their size at that time, but the variation in this respect is fully as great as seen here in its native country. One tree planted at Fulham Palace one hundred and fifty years, was only twenty-five feet high, while another growing in sandy loam at Kenwood, forty years planted, was 40 feet high with a stem nearly eighteen inches in diameter, while another at Croome, in Worcestershire, of the same age was sixty feet high. In France there are many trees of much larger size, and one at the time named, in the Botanical Garden of Avranches, twenty-nine years planted, had reached a hight of eighty-nine feet, with a stem twenty inches in diameter. I refer to these trees not as unusually large, but merely because most American authors refer to the Catalpa as a small tree, and Humphry Marshall in "The American Grove," 1785, p. 21, says the Catalpa tree " rises to the hight of about twelve to fifteen feet," or about the size of a good three-year-old specimen. The young plants usually make a late growth in the fall, especially if planted in a rich, moist soil, and the soft, unripened wood will be frequently killed, even in localities where the trees after becoming well established will seldom or never be injured by cold. A handsome ornamental tree, but the branches are liable to be broken out or split off from the main stems when the trees are planted in exposed situations.

Frank J. Scott, in his superb work "Suburban Home Grounds," in speaking of the Catalpa very truthfully says : "Though planted largely in the Northern States, and considered hardy, its beauty would be more uniform, and we should oftener see fine specimens if, when first planted, it were regarded as half hardy, and cared for accordingly." This tree is a native of Virginia, southward to Florida, and westward to Louisiana. The Catalpa is also found wild in Southern Indiana, Kentucky, Illinois, and Missouri, where in the rich bottom lands it grows to a larger size than in the South and Eastern States, but recently it has been claimed that the Catalpas found in these Western States are of a different variety or species, and the late Dr. J. H. Warder, of Ohio, gave to this western form the name of *Catalpa speciosa*. The late E. E. Barney, of Dayton,

Ohio, collected many interesting facts in relation to the value
of the Western Catalpa, which were given to the public in a
pamphlet published in 1879. The question as to whether the
Catalpa as found growing in the Western States is a distinct
species from the one found in our Eastern and Southern, is one
that can well be left to scientists to decide. My first acquaintance with the Catalpa was in Illinois, some thirty odd years
ago, and I have since seen it in all its perfection in Kentucky
and Missouri, and in those States it is certainly a grand forest
tree, and is no doubt well worthy of extended cultivation
where it will thrive as well as in its native forests. The principal points of difference claimed for their so-called Western
Catalpa, is a more erect habit of growth, larger flowers, which
appear from one to three weeks earlier in the spring. Seeds
also larger and with broader wings, and lastly the trees are
more hardy than the species, or the older and better known
Eastern Catalpa.

C. Kæmpferi.—Japan Catalpa.—A small tree resembling our
native Catalpa, with ovate, heart-shaped leaves, abruptly
pointed, sometimes three-lobed. Flowers smaller than the
American, spotted with purple. Pods and seeds smaller than
our Catalpa, and fully as hardy.

C. Bungei.—A species from Northern China, growing four or
five feet high, with handsome, dark-green leaves. Cluster of
flowers are said to be a foot long. I have had this species
growing in my grounds for many years, but it has never
bloomed, although it is apparently quite hardy.

C. bignonioldes, (var. *aurea*).—This is a handsome, golden-leaved
variety of our native Catalpa, and a handsome ornamental
tree, which with me has never been injured by the frosts of
winter.

CEANOTHUS, Linn.

A genus of some twenty-five indigenous species, all except
two are low shrubs of no special value except for ornamental
purposes, although one of the species (*C. Americanus*), indigenous to the Atlantic States has figured somewhat conspicuously
in our domestic history under the name of New Jersey Tea, as
the leaves were in early times used as a substitute for genuine
tea. All handsome little shrubs or trees with small, white or
blue flowers, in long, branching clusters. Four species are
found in the Eastern States, the others belong to the Rocky

Mountain regions, and west to the Pacific. Among the latter there are two which may be placed in the list as trees.

Ceonothus spinosus, Nutt.—Red-wood.—Leaves thick, rather rough on surface, entire, oblong, on slender stalks. Small twigs somewhat spiny. Flowers blue and very fragrant. Fruit a small drupe coated with rosin. A small tree, sometimes thirty feet high in the Coast Ranges of Southern California, where it is known as "Red-wood," from the color of the wood.

C. thyrsiflorus, Esch.—California Lilac —Leaves thick, oblong, smooth, and shiny above, somewhat downy beneath. Flowers bright blue in large, showy, compound racemes, resembling very much the flowers of the common lilac of our gardens. A tall shrub, sometimes reaching a hight of twenty feet in the Coast Ranges, from Monterey to Humboldt County, Cal.

CELTIS.—*Hackberry, Nettle-tree.*

A genus of the Nettle Family, closely allied to the Elm, but fruit a small berry-like drupe, containing only one seed. Flowers perfect or polygamous, one-petioled, singly or only a few in a cluster of a greenish color. We have some four or five species, and several natural or local varieties.

Celtis brevipes, Watson.—Leaves slightly pubescent, obliquely, ovate-oblong, pointed, an inch and a half long. Fruit about a quarter of an inch long, black. A small tree twenty or thirty feet high, and stem a foot or more in diameter. Wood soft, tough, but of little value. South-eastern Arizona.

C. Mississippiensis, Bosc.—Southern Nettle Tree.—Young leaves and twigs silky; leaves two inches long, long-ovate, pointed, sharply serrate, abruptly contracted at base; soon becoming rusty beneath. Fruit dark purple, of the size shown in figure 36, with sweet pulp, greedily eaten by several species of birds. A very large tree in the Mississippi Valley, from Kentucky southward, differing very slightly from the next.

C. occidentalis.—American Hackberry, Nettle-tree, Sugar-berry, False Elm, etc., etc.—Very similar to the last, and by some authors considered a distinct species, and by others only a northern variety. Wood soft, but difficult to split. A small tree in Vermont, and sparingly westward to Nebraska and southward, also along the Atlantic Coast in New Jersey, Long Island and southward to Florida. A rather pretty tree, seldom infested by

Fig. 36.—SOUTHERN NETTLE TREE.

insects, and the sweet fruit attracts birds in autumn. Leaves hang on late, but all drop at once when cut by frost.

Celtis Tala, Gillies—var. *pallida*, Planch.—A Mexican species, found in a few localities within the United States, Southern Florida (Garber), in the lower Rio Grande, and westward through Southern New Mexico (Botany, Mexican Boundary Survey.) A tree sometimes twenty feet high, *Celtis reticulata* of Torrey, may prove to be a distinct species, but the genus is a difficult one, and the species not readily defined, as they are quite variable; and there are many varieties. There are also two or more species and several varieties found in China and Japan.

CERCIS, Linn.—*Red-Bud, Judas-Tree.*

A genus of four species, one in Europe, one in Asia, one in the Atlantic States, and one in California, and eastward. Small trees, with smooth, heart-shaped leaves, and pea-shaped flowers, appearing early in spring, before the leaves. Seeds in small, pea-shaped pods.

Cercis Canadensis.—Red-Bud.—Leaves broadly heart-shaped, dark green, smooth, and glossy. Flowers reddish-purple, in sessile clusters, nearly covering the smaller branches in spring, before the leaves appear. A very handsome little tree, with smooth bark. In favorable situations it grows to twenty or thirty feet high. Wood hard and compact, resembling that of the apple tree, in New York, southward to Florida, and westward to Wyoming, in rich soils.

C. occidentalis.—Western, or California Red-Bud. — Leaves smaller than those of the Eastern species, and only about two inches broad; round heart-shape. Flowers rose color, and seed pods about two inches long, sharp-pointed at both ends. California, Mexico, and eastward to Texas.

C. siliquastrum.—European Judas-Tree.—Leaves somewhat kidney-shaped. Flowers larger than in our native species, but seldom produced in such abundance; a little darker in color. There are varieties with white flowers, and one with pale rose-colored flowers. A tree growing twenty feet high, a native of Southern Europe. Only occasionally hardy in our more Northern States, and the flower buds, which in all the species are formed in the autumn, are frequently killed by the severe cold of winter.

C. Japonica.—Japan Judas-Tree.—Leaves largest of all, broadly heart-shape, dark rich green, smooth. Flowers larger than those of our indigenous species, and produced in•great abundance, even from spurs on the old stems and branches. A low, much-branched shrub, six to ten feet high. Seed occasionally ripening in this country, but flower buds often killed in severe winters at the North. When this shrub was first introduced into Europe, or the United States, I have been unable to ascertain, but it must have found its way to America at a much earlier date than is usually given by our nurserymen, for in 1856, there was a large plant, eight feet high, and with many stems in the old Prince nursery, at Flushing, N. Y. This plant could not have been less than ten years old at that time, which would carry the date of introduction back to 1846, or nearly forty years ago. I obtained layered plants from this old stock in 1858, and have had it in my garden ever since. Layers of the smaller branches root very readily, and usually the first season they are put down. All the species may be propagated in the same manner, or from seeds, where they can be obtained.

Cerasus.—(Cherry).—See *Prunus.*

CHILOPSIS, Don.—*Desert Willow.*

A genus represented by only one species in the United States. It is closely allied to the Catalpa-tree, and our common Trumpet Creeper (*Tecoma radicans*). A large order in the tropics, but has few representatives in the temperate zone.

Chilopsis saligna, Don.—Texas Flowering Willow, Desert Willow. —Leaves long, narrow, or linear-lanceolate, four to six inches long, opposite in whorls or irregularly alternate, entire and slightly sticky when old. Flowers funnel-form, one to two inches long, white and purplish, in short, terminal racemes. A small but handsome tree, twenty feet high. In Southern Texas, in Mexico, and Southern California. Probably not hardy anywhere in our Northern States, but a valuable ornamental shrub or tree for the South. P. J. Berckmans, of the Fruit-land Nurseries, Augusta, Ga., writes me that it grows freely from cuttings, making plants five or six feet high the first season.

CHIONANTHUS.— *White Fringe Tree.*

A genus closely related to the common ash (Fraxinus). Only one species, and this muchadmired for its long, loose panicles of snow-white flowers, resembling cut paper. Appearing late in spring.

Chionanthus Virginica.—White Fringe.—Leaves entire, oval or obovate, three to five inches long, the lower surface sometimes rather downy. The long, slender petals of the flower an inch long. Fruit, a single-seeded drupe, purple, with bloom. There is a variety of the White Fringe Tree, formerly known in nurseries as *C.* var. *angustifolia.* Leaves long and narrow, and flowers usually larger and more abundant. Of late years this variety appears to be scarce in nurseries, and is seldom mentioned in catalogues. I do not now call to mind but one specimen of this variety, and this is or was a few years ago growing in the grounds of Mr. Charles Downing, Newburgh, N. Y. There may be others scattered about the country, and it is well worthy of preservation. The White Fringe is a large shrub, growing twenty to thirty feet high in the South. It may be budded or grafted upon the ash, and when worked on such stocks, grow faster and make better shaped trees than when growing on their own roots. Native of Southern Ohio, Pennsylvania, and thence southward to the Gulf of Mexico.

CHRYSOPHYLLUM, Linn.—*Star Apple.*

Handsome trees of tropical regions, usually with leaves bright green above, and of a bright golden or coffee-color beneath. Fruit fleshy, and of most species edible. Two species are found within the United States, on the southern border, viz.: *C. microphyllum*, DC., in Southern Florida, where it has probably been introduced from the West Indies, *C. Oliviforme*, Lam., in same localities, fruit black when ripe, scarcely edible. Both small trees of no value outside of the tropical climates.

CLADRASTIS, Raf.—*Yellow Wood.*

Trees with white, pea-shaped flowers in long drooping racemes, succeeded by small, flat pods, containing a number of small, brown seeds. Only one indigenous species.

Cladrastis tinctoria, (*Virgilia lutea* of Michaux).—Leaflets seven to eleven, oval or ovate, three to four inches long. Flowers white, fragrant, an inch long, in large, drooping panicles, late in spring. Bark smooth, of a grayish-color. One of our cleanest and most beautiful forest trees, usually with an open, rather wide spreading head. Wood light yellow, very firm and hard, makes excellent fuel, and is desirable for all kinds of mechanical purposes, where a fine, hard wood is required. The roots are almost as tough and strong as a hempen rope, and it requires a sharp spade to cut them in transplanting. I am reminded of this characteristic of the roots, from the fact that about twenty-five years ago I sent to Kentucky, where this tree is most abundant, and secured two pounds of the seed, from which I raised several thousand trees. After transplanting them once or twice, and they had reached a hight of three or four feet, I offered them free to my neighbors, any one who would accept and take up few or many; but as very few persons were acquainted with the trees, I only succeeded in distributing a small number, and the remainder were dug up and burned. One of this lot is now growing on my lawn, and it is a noble specimen, and not excelled by any tree in my grounds. When planted in a forest, it will grow tall and straight. It deserves to be far better known and more extensively cultivated than it has been. Native of Kentucky and Tennessee, but not abundant.

C. amurensis, Bentham and Hooker.—This is an Oriental species, which has been introduced into this country, and is a free grower and quite hardy.

CLIFTONIA, Banks.—*Buckwheat Tree.*

A small tree or shrub, the *Mylocarpum,* of Willdenow. Only one species, the

Cliftonia ligustrina.—Buckwheat Tree.—Leaves evergreen, oblong, smooth, and somewhat glaucous. Flowers white, fragrant, in racemes two to four inches long, appearing in March and April. A tree sometimes twenty feet high, along the borders of streams in Georgia, Alabama, and Florida. Propagated from seed, layers, or green cuttings, in a propagating house.

CLUSIA.—*Balsam Tree.*

A genus of tropical trees or shrubs with resinous yellow juice and rough evergreen opposite leaves. Fruit of some of the species edible, only one native of the United States.

Clusia flava, Linn.—Balsam Tree.—Leaves with short stalks obovate, finely veined. Flowers polygamous, single, or in threes, on short axillary or terminal peduncles, yellow. Fruit pear-shaped, containing about a dozen seeds, imbedded in a soft pulp. A small tree about thirty feet high in Southern Florida, and the West Indies.

COCCOLOBA, Jacq.—*Pigeon Berry.*

The species of this genus are principally large evergreen trees, a few only large shrubs, mainly native of the West Indies, but a few extending into South America. All have greenish flowers in axillary and terminal racemes. Over twenty species are described in botanical works, two are found in Southern Florida.

Coccoloba Floridana, Meisner.—Pigeon Plum.—Leaves smooth, elliptical, obtuse at each end; two to three inches long. Flowers in slender racemes, either terminal or on short lateral branches. Fruit a rather pleasant flavored berry. A tree about forty feet high, with hard, firm wood. In rocky situations in the West Indies, and at Key West in Florida.

C. unifera, Jacq.—Sea-side Grape.—Leaves smooth, with short stem; roundish heart-shaped. Flowers in terminal erect racemes, sometimes a foot long. Bark on branches smooth, but on the stems becomes rough and furrowed. Fruit in long clusters, the size of large cherries, but pear-shaped, purple when ripe, with thin pulp of a sub-acid taste. Often referred to by travellers and residents of Florida as an excellent dessert fruit. A large tree, forty to sixty feet high, and stem two or more feet in diameter. Wood heavy and hard, violet-colored, highly valued for cabinet work. An astringent extract of the wood known under the name of *Kino*, is sometimes used as a medicine. Southern Florida and the West Indies. In the Bahamas it is called the Mangrove Grape Tree.

CONDALIA, Cavan.—*Bluewood, Logwood.*

A genus of three species, one in South America, and two in the United States. Small evergreen shrubs, but one, the *Condalia obovata*, sometimes a small tree in Eastern Texas and westward to Arizona. Closely allied and resembles the common Buckthorn, *Rhamnus Caroliniana*. Wood heavy, but too small to be of much value.

CONOCARPUS, Linn.

A small genus of evergreen tropical shrubs, the bark of some of the species being employed in tanning. The fruit resembles a pine cone in form. We have one native species.

Conocarpus erecta, Jacq.—Button Tree.—Leaves smooth, oblong or lanceolate-pointed. Flowers on short, spreading pedicels.

Fruit in an ovoid imbricated cone. A small tree fiom Tampa, Florida, southward to Brazil, in sandy soils.

CORDIA, Linn, Plumier.

Sub-tropical or tropical trees, and shrubs of the East and West Indies, and other warm climates. Highly valued as ornamental trees. The greater portion of the species are American. Two of the four found in the United States reach a hight of twenty or more feet.

Cordia Borissieri, A. DC. Leaves oval, or oblong-ovate, soft, woolly, but becoming smooth or rugose when old. Flowers white, with yellow center; five-lobed, and rather downy on outside. Two to four one-celled hard seeds, enclosed in a small, pulpy fruit. A small tree, twenty feet high, along the southern border of Texas and New Mexico, and westward.

C. Sebestena, Linn. — Rough-leaved Cordia. — Leaves large, four to eight inches long, ovate-oblong, rough to the touch. Flowers are deep yellow-orange, in large, terminal corymbose racemes. Fruit is a round, or pear-shaped drupe, containing a deeply furrowed nut. A handsome, ornamental tree in South Florida, and the West Indies. The botanist, Catesby, states that the wood of this species is of a dark brown, approaching to black; very heavy, and containing a gum, in smell and appearance resembling that of Aloes. In the Bahama Islands is called Lignum Aloes (Nuttall's North American Sylva, Vol. II. p. 146.)

CORNUS, Tour.—*Dogwood.*

A large genus, principally shrubs and small trees, a few perennial herbs, mostly of the northern hemisphere; one in South America. Fifteen of the twenty-five species, known to botanists, are indigenous to the United States, but only two reach the hight of twenty feet or more. Flowers perfect, small, in compact clusters or heads, usually quite inconspicuous, but in some species they are enclosed at first in a corolla-like involucre, which, upon expanding, is very showy. This flower-like envelop is usually referred to as the flower of the common Dogwood of this country.

Cornus florida, Linn.—Flowering Dogwood.—Leaves ovate, pointed, smooth on both sides. Flowers small, greenish-white, enclosed in a large four-leaved, white involucre, sometimes tinged with red. Fruit an oval, bright red drupe, with an agreeable tasted pulp, much sought by several species of birds.

The flowers appear in spring, before the leaves, making this tree one of the most strikingly beautiful to be found in our forests. If it was not so common, it would be more highly prized and more extensively planted as an ornamental tree than any tree ever introduced from abroad. Wood very hard and close-grained, and on this account often called American box-wood ; valuable for the handles of small mechanical implements. A tree sometimes forty feet high, with broad, roundish head, and a stem nearly or quite one foot in diameter. Common on high, dry ground, from Canada to Florida, and westward to Texas. A handsome weeping variety of this species which has been named *C. florida pendula*, was found near Warren, Baltimore County, Md., by Dr. W. S. Thompson, who sold it to Thos. Meehan, of the Germantown Nurseries, who, after propagating it for a few years, sold the entire stock to a gentleman in Rochester, N. Y., from whom it passed into the hands of Pratt Brothers, of the same city. This is a beautiful weeping tree, with flowers of the same size and color as the species.

C. Nuttalli, Audubon.—Nuttall's Dogwood.—Leaves more or less pubescent obovate, three to five inches long, pointed at both ends ; involucre of from four to six oblong bracts, sometimes three inches long, yellowish or pure white, often tinged with red. Flowers numerous in large heads. Fruit a little larger than in the last, of a bright crimson color. This species resembles the eastern dogwood very closely, and may be only a western variety. A small tree in some localities, in others fifty to seventy feet high. Wood same as that of *C. florida*. On the Pacific Coast, from Monterey, Cal., northward to the Fraser River. Four other species are found in the Rocky Mountains and on the Pacific Slope, viz.: *C. sessilis*, Torr. *C. Californica*, Meyer. *C. pubescens*, Nutt. *C. glabrata*, Benth., and *C. Torreyi*, Watson, all shrubs, from five to fifteen feet high. Of the Eastern species of the Cornus, there are about a half dozen different species, all low shrubs, except one, and this is a small herbaceous plant. Of European species, none grow large enough to be classed as trees, and only one or two are cultivated for ornament or other uses. The Cornelian Cherry, *Cornus mascula*, is a large shrub, bearing yellow flowers in spring, succeeded by handsome oblong fruit, of a bright red color, edible, and sometimes used as a substitute for olives.

The Turks value the fruit highly for flavoring sherbet, and it is considered useful in dysentery. There is a variety with bright yellow fruit, of the same size and flavor as the red. Both grow and fruit freely in our Northern States. Another variety has variegated leaves. The above and many other varieties and species are cultivated in nurseries, descriptions of which can be found in nurserymen's catalogues.

CRATÆGUS.—*Hawthorn.*

A genus of small trees or shrubs, about a dozen species indigenous to the United States, all bearing small, white flowers in spring and early summer. Wood hard, resembling that of the common apple tree, but seldom large enough to be serviceable, except for handles for small tools and similar purposes. The early settlers in our Eastern States thought they had found in our native hawthorns a plant equal, or superior to the English hawthorn for hedges, and while they are in every respect equal if not superior to the exotic species for such purposes, they have so many insect enemies in this country, that hedges made of them are liable to be destroyed before or soon after reaching a useful size. The common apple tree borers attack the hawthorns, and when set in close hedges the destruction of these insects, hidden as they are, in the stems and branches, is impracticable, if not entirely impossible. For this reason, and this only, the hawthorns are not to be recommended for hedges, especially in our Northern States, where several different species of what are commonly termed apple-tree borers abound. The hawthorns are mostly thorny, with simple or lobed leaves. A few of the species are evergreens. Seeds of the hawthorns are very hard, and do not usually germinate until the second season after planting. The best way to manage these seeds is to mix them in sand or soil, after the pulp is removed, and then bury them on the north side of some building, or in a shady place, and leave them undisturbed until the opening of the second year after gathering, then take out and sow in drills. The seed should not be allowed to get very dry before being put into the sand.

Cratægus æstivalis, Torr. and Gray.—Summer Hawthorn.—Leaves spatulate or wedge-obovate, somewhat rough, crenate above the middle. Flowers solitary, or only three to five in a cluster. Fruit large, round, red, of a mild acid flavor. Ripe in summer. A small tree, but sometimes more than thirty feet high. From Virginia southward in the pine-barrens, and along the margin of streams and ponds.

C. apiifolia, Michx.—Parsley-leaved Thorn.—Leaves and young branches white, pubescent or downy, when young ; leaves small, divided into five to seven lobes, sharply toothed. Fruit round, red when ripe, and edible. A small tree in moist soils, from Virginia southward.

C. arborescens, Elliott.—Tall Hawthorn.—Leaves smooth, thin, oval or elliptical, sharp at both ends, and finely serrate. Flowers abundant in large clusters. Fruit ovoid, red. Branches with a few large stout thorns. Small tree, sometimes thirty feet high. Georgia, Florida, and west to Texas.

C. berberifolia, Torr. and Gray.—Barberry-leaved Hawthorn.— A doubtful species, about which little is known, although mentioned in several botanical works. Said to be found in Louisiana, and grows thirty feet high.

C. coccinea, L.—Scarlet-fruited Thorn.—Leaves bright green, smooth, thin, roundish-ovate, deeply cut or lobed, on slender petioles. Flowers large, in large clusters. Fruit small, round, · or pear-shaped, bright red, scarcely eatable. A variable species of which several varieties are mentioned in botanical works. A small tree, seldom over twenty feet high. Common in Canada and nearly all of our Northern States, and southward to Florida.

C. cordata, Ait.—Washington Thorn.—Leaves large, thin, bright green, broadly heart-shaped, sometimes almost triangular, often three to five-cleft or lobed. Thorns numerous, but slender. Fruit small, round, not larger than a good-sized pea, red. A small tree, formerly highly recommended for hedges. Virginia, Kentucky, and southward.

C. Crus-galli, L.—Cock-spur Thorn.—Leaves smooth, thick, oblong-ovate, finely serrate above the middle, narrowed at the base into a slender petiole. Flowers large and numerous. Fruit large, bright red. Thorns numerous, and very long and sharp. There are many natural or wild varieties. A tree fifteen to twenty-five feet high, in both Northern and Southern States, and westward beyond the Mississippi.

C. Douglassii, Lindley.—Douglass Thorn.—Leaves broadly-ovate, usually somewhat lobed or cut above, and rather finely serrate, one to three inches long, with short stem. Flowers large and numerous. Fruit a half inch in diameter, dark purple, sweet and edible. A small tree in California, and northward to Puget Sound, along streams.

C. flava, Ait.—Yellow or Summer Hawthorn.—Leaves thick,

wedge-obovate serrate, and slightly lobed above the middle, tapering into a short petiole. Flowers in small clusters or solitary. Fruit quite large, round, or pear-shaped, yellow or greenish yellow, sometimes tinged or spotted with red, and pleasant flavored. A tree twenty feet high, from Virginia, south and west.

C. rivularis, Nutt.—River Hawthorn.—Leaves ovate, or oblong-ovate, contracted at the base into a short, slender petiole, irregular serrate, but rarely divided or lobed. Flowers small, few in a cluster. Fruit small, black, and of insipid taste. A small tree, seldom twenty feet high. California and northward, and east to Montana.

C. spathulata, Michx.—Spatula-leaved Thorn.—Leaves small, spatulate or broadest above, narrowing at the base, those on the young, downy branches, somewhat cut or lobed. Flowers in large clusters. Fruit very small, red. A small tree, scarcely twenty feet high. Virginia and southward.

C. subvillosa, Schrader.—Soft-leaved Thorn.—Leaves roundish, soft, downy, not tapering, but often heart-shaped and double-toothed. Flowers large and abundant. Fruit about a half inch in diameter, dull red and of an insipid flavor. Western States and South. A small tree.

C. tomentosa, L.—Black or Pear Hawthorn.—Leaves soft, downy when young, becoming smooth with age, three to five inches long, oval or obovate, and but slightly lobed. Flowers large, often an inch broad. Fruit very large, nearly or quite an inch long, red or orange yellow, very variable in flavor, sometimes sweet, and that of other trees sprightly sub-acid. I have eaten varieties of this fruit in Western New York, also in Wisconsin, and other localities in the West that were really delicious. There are many and widely variable natural varieties. A shrub or tree of thirty feet high. Vermont, westward to Wisconsin and Iowa, and southward to Georgia.

C. parvifolia, Ait.—Small-leaved Thorn.—Leaves only about an inch long, obovate-serrate, with very short stalk; spines numerous, long and slender. Flowers mostly solitary Fruit large, round or pear-shaped, greenish-yellow. A small shrub four to six feet high, in pine-barrens from New York southward.

There are many exotic species, all with one exception are natives of Northern Asia, Europe, and North America. The

English Hawthorn (*C. oxycantha*), was introduced and planted
here quite extensively by the English colonists, and many attempts were made to establish hedges of this thorn, but it was
soon discovered that when close pruned, as in Great Britian,
the leaves and young twigs were scorched by the hot sun in
summer, and hedges of this kind were pronounced a failure.
Still every few years there would be a revival of the interest in
the English Hawthorn for hedges, and others would try it, and
for a few years the plants would appear to be doing well, but
close pruning, drouths, and insects usually combined as
enemies of such hedges, but even at this time the remnants of
those old hedges are to be seen in our Eastern States. But
there is no difficulty in raising good specimen plants of this
species, when planted singly and in a moderately moist and
rich soil. There are a large number of ornamental varieties described in nurserymen's catalogues, and cultivated in gardens.
There are varieties with single white, pink, and crimson
flowers. Also those with double flowers of the same colors,
all really beautiful little trees, well worthy of a place in every
garden where there is room to plant them. The varieties are
propagated by budding and grafting upon stocks of the common European or native species.

The Evergreen Thorn (*C. Pyracantha*), of Southern Europe,
has become naturalized and run wild in some of the Middle
and Southern States, and in the latter is considered a most excellent hedge plant, as it is a compact growing, very spiny
shrub, with shining, evergreen leaves. White flowers in spring,
succeeded by clusters of small orange colored berries. There
is also a variety with white berries. The evergreen thorns are
hardy in our Northern States, but the leaves turn brown or
are entirely destroyed in winter, for this reason they are not
popular in cool climates.

CYRILLA, Linn.

A genus of two species of small evergreen trees, with opposite leaves, and white, five-petaled flowers, in terminal or
lateral racemes. Only one species found in the United States.

Cyrilla racemiflora, Walt.—Leaves oblong, three or four inches
long, on very short petioles. Flowers numerous, in long,
straight racemes. Fruit a round one or two-seeded drupe,
tipped with a conspicuous, thick or slender point. A small
tree, sometimes thirty feet high, along the shady banks of
streams and ponds, from North Carolina to Florida.

DIOSPYROS.—*Persimmon*.

A large genus of handsome trees of the Ebony Family. Flowers polygamous or diœcious, the fertile ones solitary, and the sterile smaller and in clusters. Represented in the United States by two species.

Diospyros Virginiana, Linn.—Persimmon.—Leaves ovate-oblong, smooth, dark-green, with very short peduncles. Flowers pale yellow or greenish, four-cleft. Fruit large, plum-like, containing four to eight seeds. The fruit is quite variable in size, shape, and quality. The most common form is round, but occasional varieties may be found with oblong or oval fruit. The taste of all is intensely acerb or astringent while green, but becoming eatable, and in some instances of excellent flavor, when ripe. In the Northern States, the persimmon seldom ripens until touched with frost, still there are varieties which ripen earlier, and are fully mature a month before the arrival of frosts. The persimmon is a fruit well worthy of the attention of our pomologists, and could, no doubt, be greatly improved by cultivation, and new varieties produced equal to that of any species of the same genus found in other parts of the world. An occasional variety is found with almost seedless fruit, or at most containing only one or two seeds. Improved varieties may be readily multiplied by budding or grafting upon seedlings of the wild trees. A handsome, ornamental tree with clean, bright foliage, and very heavy, close-grained, dark-brown wood. In rich soils, this tree sometimes reaches a hight of sixty to seventy feet. In Southern Connecticut, Northern New Jersey, and southward to Florida. Also abundant in Ohio, Michigan, Illinois, and Iowa, and southward.

D. Texana, Scheele.—Mexican Persimmon.—Leaves cuneate-oblong, round at the apex, and only an inch or two long and somewhat downy. Flowers silky, tomentose on the outside. Fruit downy when young, round, black when matured, containing three to eight triangular seeds. Fruit of excellent quality, said to ripen in August. A small tree twenty to thirty feet high, with white, but heavy wood. Southern and Western Texas, and in Mexico. The species are pretty widely distributed over the world, in Europe, Asia, New Holland, the East and West Indies, and several of the larger Islands furnish one or more. The Date-Plum of Europe (*D. Lotus*), has furnished material for many of our ancient romances. and quite recently we have discovered that the Japanese have, and cultivate many varieties of the *Diospyros Kaki*, some of them have remarkably large and fine fruit. These Japan varieties are being quite successfully cultivated in California, and by a few persons in all our Southern States, but they have not proved quite hardy north of Washington. Specimens of the Japanese persimmons, planted in the neighborhood of New York City, have in some instances escaped being killed by cold, but they are by no means hardy in our Northern States. The introduction of their fruit will no doubt awaken an interest in this very deserving genus, and may result in the improvement of our native varieties.

DIPHOLIS, A. DC.

A genus of tropical or sub-tropical trees, closely allied to the Burrelias referred to on a preceding page. One only extending as far North as Southern Florida.

Dipholis sallcifolia, A. DC.—Willow-leaved Dipholis.—Leaves evergreen, oblong-lanceolate, pointed, smooth, tapering below into a petiole. Flowers small, oval, one-seeded berry, the size of a pea, the pulp being very dry, and almost destitute of juice. A large tree fifty to sixty feet high on the Florida Keys.

DRYPETES, Vahl.

Tropical evergreen trees or shrubs, with entire smooth, petioled leaves, and minute diœcious flowers without petals. Fruit a one or two-seeded drupe.

Drypetes crocea, Poit.—Leaves three to four inches long, oblong, pointed at both ends, and of a leathery texture. Flowers small, greenish-white, many in a cluster. Fruit somewhat four-angled, and velvety, containing one seed. Usually considered only a shrub, but some authorities report it a large tree in Southern Flordia and in the West Indies. Another but smaller species the *D. glauca* is credited to the same region by Dr. Chapman in his Flora of the Southern States, on the authority of Dr. Blodgett. But this may be only a variety of the last.

EHRETIA, Linn.

A genus of tropical evergreen trees or shrubs, of about a dozen species in the East and West Indies, and New Holland. These have small, white flowers and edible fruit. We have one species.

Ehretla elliptica. DC.—Leaves smooth, oval or oblong, sometimes serrate. Young branchlets and flowers hairy or downy.
Flowers white and fragrant. Fruit about the size of peas,
round and yellow when ripe, containing one seed, furrowed on
the back. A small tree twenty to thirty feet high in South
Florida, and westward in the swamps of Southern Texas and
Mexico. For *E. radula*, Poir., see *Bourreria Havanensis*.

EUGENIA, Micheli.—*Allspice.*

A large genus of evergreen, tropical or sub-tropical shrubs or
small trees belonging to the Myrtle Family. The greater part
are indigenous to the Caribbean Islands. Flowers small, and
the fruit a globose berry, crowned with a persistent calyx.
Seeds one or two, roundish and large. Four of the species are
found on our southernmost border.

Eugenia buxifolia, Willd.—Box-leaved Allspice.–Leaves smooth,
rather thick, obovate-oblong, about an inch and a half long.
Flowers minute, with reddish petals. Berry black, about
three-eighths of an inch in diameter, containing one to three
seeds. Wood hard and close-grained. A tree about twenty
feet high, growing in sterile places, near the sea, at Key West,
and on several of the islands in the West Indies.

E. dichotoma, DC.—Small-leaved Eugenia.—Leaves oblong-
obovate, roughened with appressed hairs, becoming smooth.
Flowers long, peduncles reddish, berry about the size of a pea,
one-seeded. Branches covered with a smooth, grayish bark.
Wood hard. A small tree in Southern Florida and the West
Indies. Cultivated in conservatories at the North on account
of its fragrance.

E. procera, Poir.—Tall Eugenia.—Leaves smooth, ovate-taper-
ing, but with a rather blunt point. Flowers solitary, or only
two to four together, same color as the last. Berry small,
round, not larger than a grain of black pepper ; one-seeded.
Wood white, close-grained. The bark on the twigs and smaller
branches. silvery white. A small tree, twenty to thirty feet
high at Key West and in the West Indies.

FAGUS, Tour.—*Beech.*

Mostly large, handsome timber trees, those of the northern
hemisphere deciduous, but there are two evergreen species in
South America, and one in New Zealand. Sterile flowers in
small heads, on drooping peduncles, the fertile ones in pairs on

the summit of a scaly-bracted peduncle. The fruit, a pair of triangular-shaped nuts, enclosed in a prickly involucre.

Fagus ferruginea, Ait.—American Beech.—Leaves oblong-ovate, silky on both sides when young, becoming smooth with age, except on the veins; edges serrate or distinctly toothed. Nuts well known for their rich and delicate flavor. The beech is one of our most noble and valuable forest trees, and of a graceful habit. The leaves remain green until quite late in the season, seldom changing color until cut by frost, when unfortunately they do not drop at once, but remain attached to the branches for weeks or months, a few dropping at a time, all through the winter. This habit of retaining the dead and dry leaves in winter is an objection to the beech as a lawn tree, because there is a constant littering of the grounds until the new foliage pushes in spring. Bark on the stem and branches smooth, and of a grayish-white color, in fact, the beech may be termed one of the cleanest-looking trees of our forest, and it is seldom attacked by insects. Wood very hard and firm, susceptible of a very fine polish, and is next to the hickory in value for fuel. In some soils, the wood is white even in quite large trees, but in others it is of a rich brown or reddish, and I have seen trees in our northern woods that were over two feet in diameter that did not have more than two inches of white wood on the outside, all the rest being of a brown color. Wood extensively employed for making plane stocks, handles for tools, cabinet work, hewn timber, and other purposes. The roots of the beech do not usually go deeply in the ground, but keep close to the surface, especially in moist, stony soils, which it frequents. It is an excellent tree for planting in rocky, exposed situations, as its slender, tough branches withstand high winds and cold storms. The nuts may be treated the same as recommended for the chestnut, but almost any quantities of the natural seedlings can be procured in our northern woods, and they can be readily transplanted without much loss. A common tree in Nova Scotia, Canada, and all of our Northern States and southward, along the mountains and valleys in rich soils to Florida.

The European Beech (*F. sylvatica*) is very similar to the American, but distinguished by shorter and broader leaves, with somewhat wavy margins. Of this species there are many varieties in cultivation, and among the best known are the following, all of which are large trees : Purple-leaved (*F.* var. *pur-*

7

purea), found in a forest in Germany. Crested-leaved (*F.* var. *crestata*). Copper-leaved (*F.* var. *cuprea*). Silver-leaved (*F.* var. *fol argentea*). Golden-leaved (*F.* var. *fol aurea*). Cut-leaved (*F.* var. *incisa*). Fern-leaved (*F.* var. *heterophylla*). Oak-leaved (*F.* var. *quercifolia*). Weeping Beech (*F.* var. *pendula*). These are all hardy in our Northern States, and are propagated by grafting on stocks of the common American or European beech.

The *F. antarctica* is a deciduous species, native of the region about the Straits of Magellan and Patagonia, S. A., where it grows to a large tree sixty or more feet in hight. Leaves small, and fruit not much larger than buckwheat. *F. betuloides* is indigenous to the same region, but has smaller and evergreen leaves. A small tree, twenty to thirty feet high. I do not know that any attempts have been made to introduce these species, but it is likely that they would succeed in the United States.

Fagus obliqua is another evergreen species, native of the Andes, S. A., and *F. Cunninghami* is a large evergreen tree in New Zealand.

FICUS, Tour.—*Fig.*

This is an extensive genus of evergreen trees and shrubs of the easiest cultivation, all native of tropical or subtropical countries, extending entirely around the world. The *Ficus elastica* is the well-known India rubber tree, and the celebrated Banyan Tree, of India, is the *Ficus Indica*, the juice of which is used by the Hindoos to cure the toothache, and the bark is considered an excellent tonic. *Ficus carica* furnishes the well-known figs of commerce. There is nearly or quite one hundred species, of which three are natives of the United States.

Ficus aurea, Nutt.—Small-fruited Fig.—Leaves smooth, oblong, entire, narrowed, but rather blunt at both ends. The figs are about the size of peas, produced in pairs, close to the stem, and of an orange-yellow color when ripe. According to Dr. Blodgett, the discoverer of this species, it is parasitical on other trees, but by destroying its supporters it at length reaches the ground, and then takes root in the earth, and becomes a large tree. Key West, Florida.

F. brevifolia, Nutt.—Short-leaved Fig.—Leaves about two inches long, by one and a half broad. Figs about the size of small cherries, on the ends of the twigs, light purple or red when ripe. A small and rare tree, at Key West, Florida.

F. pedunculata.—Cherry Fig.—Leaves ovate or oval, rather thick, of a leathery appearance on upper surface, somewhat heart-shaped at base, stalk slender. Fruit nearly round, about the size of small cherries, greenish-yellow or purplish when ripe. A lofty tree of fifty feet high, sending down aërial roots, like the famous Banyan tree. Southern Florida and the West Indies.

FRAXINUS, Tour.—*The Ash.*

A very extensive genus, and most of the species are large trees, well adapted for planting in forests. The leaves are odd-pinnate, and from five to nine leaflets. Flowers diœcious or mostly so, very small, not at all showy, except on a few species, and those native of our Eastern States, are entirely destitute of petals. Fruit winged, sometimes only above, in others all around, and in a few the seed is three-winged and three-celled. Trees usually well furnished with small, fibrous roots, and not difficult to transplant and make grow. Quite free from the habit of producing suckers, when the roots are broken or otherwise injured.

F. AMERICANA.—*White Ash.*

Leaflets ovate-oblong, or lanceolate-oblong, pointed, edges nearly or quite entire, smooth on the upper surface, and downy beneath. Fruit rather short, somewhat wedged-shape, rounded at lower end, winged above. A very large, handsome tree, with gray furrowed bark on the main stem, and that on the young branchlets of a greenish-gray color, smooth, buds rusty-colored in winter. This species deserves special attention on account of the great value of its timber, it being one of the toughest and hardest of the whole genus. The wood of what is termed second growth trees, or those springing up after the original forests have been removed, or from seed, scattered in open fields, is usually superior in toughness to the first growth or large trees. The superiority of many of our farm implements is due in a great measure to the tough, but light ash timber, which enters into their construction. It is not only used for agricultural implements, but for carriages, oars, cabinet work, floors in dwellings, in fact, white ash is well adapted to all purposes where a light colored, tough, and hard wood is wanted. A tree of rapid growth, and reaching a hight of seventy to eighty feet, and thrives in a great variety of soils, but succeed best in a rich, moist one. Common in all our Eastern States, Canada, and westward to Nebraska, and southward to the Gulf, but it is becoming scarce in many localities, where a few years ago it was quite abundant. We have no native tree more worthy of extended cultivation than the White Ash, or one the wood of which is likely to be in greater demand a few years hence.

F. A. var. *microcarpa*, is a kind of a sport, found in Alabama, with very small fruit, but seedless or without any germ or meat, as usually termed.

F. A. var. *Texensis*, is a small tree with about five leaflets, with fruit scarcely an inch long. Has been described under various names by different botanists. A low growing tree, on rocky hills, from Austin, Texas, to the Rio Grande. Gray, in Flora of North America.

F. anomala, Torr.—Leaves mostly simple, sometimes two or three foliate ; oval or heart-shaped, thin ; the young shoots soft, pubescent. Flowers in short panicles. Fruit oblong, wings extending to the base. A small tree in Southern Utah, sometimes twenty feet high or over in the canyons.

F. cuspidata, Torr.—Leaves five to seven, lanceolate, and gradually tapering to a sharp point. Flowers with a four-pointed corolla, a half inch long. Fruit small, about a half inch long. A small shrub, six to eight feet high in Southwestern Texas and New Mexico.

F. dipetela, Hook. and Arn.—Two-petaled Ash.—Leaflets five to nine, oval or oblong, serrate, and only an inch or two long. Flowers with two petals, in panicles clustered on short, lateral spurs. Fruit about an inch long, broad at the top, with sharp edges below. This is the *Ornus dipetela* of Nuttall. A small tree in Western California, and of which there are two or three natural varieties.

F. Greggii, Gray.—Is a small shrub, closely allied to the last, and found in limestone soils in Southwest Texas and in New Mexico.

F. Oregona, Nutt.—Oregon Ash. — Leaflets five to seven, lanceolate-oblong to oval, entire, or nearly so, two to four inches long. Fruit about an inch long, somewhat club-shaped, widening upward into a long, broad wing. A large tree, with wood resembling the White Ash of the East. Along streams in Washington Territory, and southward near the coast to San Francisco.

F. pistaciæfolia, Torr.—Leaflets five to nine, petioles short, from lanceolate to oval, entire, or slightly serrate. Fruit small and crowded, somewhat club-shaped. A small tree, but rather stocky ; twenty to thirty feet high, with stems a foot or more in diameter. Southwest Texas, Arizona, and Mexico. A rigid form of this (var. *coriacea*), with thick leathery leaves, frequents the arid districts of Arizona.

F. platycarpa, Michx.—Water Ash.—Leaflets five to seven, ovate-oblong, pointed, and sharply serrate or entire. Fruit elliptical, broad above, two inches long, contracted below, sometimes three-winged. Young branches round, smooth or pubescent. A small tree, thirty or forty feet high, from Southern Virginia in swamps to Florida, and westward to Louisiana. Said to be also found in Cuba.

F. pubescens, Lam.—Red Ash.—Leaflets seven to nine, oblong-ovate, gradually pointed, green above, and pale velvety beneath. The young branches and leaves are quite velvety at first, hence the specific name. Inside of the bark reddish or cinnamon-color. A small, but rather slender tree, in swamps, and along streams. Canada to Florida, and westward to Dakota, but far more common in the Eastern than in the Western States.

F. sambucifolia.—Black Ash.— Leaflets seven to eleven, green on both sides, oblong-lanceolate, form a roundish base, gradually tapering to a point, finely and sharply serrate. Flowers entirely naked. The crushed leaves exhale the odor of the common Elder (*Sambucus*). Fruit flat, and winged all around. Tree of moderate size, but quite tall, stems slender, a foot in diameter in trees sixty to seventy feet high, in rich swamps, and along streams. Wood coarse-grained, but exceedingly tough, readily separated into thin layers, hence its extensive use by the early settlers in our Northern States for seating chairs, making baskets, and various other household uses. A valuable tree for planting in moist and wet soils. Canada to Virginia, and westward to Arkansas.

F. viridis, Michx.—Green Ash.—Leaflets five to nine, bright green on both sides, sometimes a little whitish on the under side ; oblong-ovate, more or less toothed. Fruit flattish, two-edged at base, widening into a long, lance linear wing. A variable species, closely allied to the Red Ash, but a smaller tree, most common in low grounds. West and South, but found from Canada to Florida, and westward to Dakota and Arizona.

Several ornamental varieties of the American Ash have been propagated by our nurserymen, and among them the Aucuba-leaved, Walnut-leaved, and Cloth-like-leaved are perhaps the best.

FOREIGN SPECIES AND VARIETIES.

The foreign species and varieties of the Ash, are so very numerous that my limited space will not admit of mentioning only a few of the most distinct and desirable. The common European Ash (*Fraxinus excelsior*), has been cultivated so many centuries, and so extensively that it has yielded a very large number of interesting varieties, that have been perpetuated by the usual methods of budding and grafting *F. ex.* var. *pendula* (Weeping Ash) should be grafted high in order to allow of the branches descending a considerable distance before touching the ground. *F. ex. aurea pendula*, similar to the last, but with golden yellow bark. *F. ex.* var. *aurea*, is of upright growth, but with golden bark. *F. ex. salicifolia* (Willow-leaved Ash). Leaves narrow, resembling the Willow, and there is another closely allied variety, known as *F. scolopendrifolia* which has long, drooping, grass-like foliage. *F.* var. *monophylla* has single, entire-leaves, instead of pinnate foilage, which is the usual form of the Ash leaf. The Flowering Ash (*F. Ornus*) of Europe is a handsome tree, with large clusters of white flowers. Hardy, and should be more frequently planted in the gardens and parks of this country. The varieties of Ash are not always constant, and often require pruning away of branches which revert to the original type. This is especially the case with the Weeping, and Variegated-leaved varieties. Some-times branches of the Weeping Ash assume an erect habit, and if not removed, will seriously interfere with the growth of the pendulous ones. In fact, these trees are all more or less in-clined to " sport," and as an instance of the sudden and wide departure from the normal form, I will call the readers at-tention to the Remilly Ash, shown in figure 37, copied from "The Garden," Eng. This handsome Weeping Ash tree is grow-ing at Remilly, not far from Metz, France, and has been named by Mr. Carriere, *Fraxinus pendula remillyensis*. This tree is about sixty-five high, with a stem nearly six feet in circumference. The leaflets are broad, and very much like those of the common Weeping Ash. A few years ago three shoots started from the pendulous branches as shown at *A,B,C,* these growing upright, and the leaves on these are very narrow, taper-pointed, and of a deeper and more glossy green. If such freaks of nature occur in such large old trees, we may

Fig. 37.—REMILLY ASH.

confidently expect equally as curious ones to occur among seed-lings raised under artificial conditions.

Several species and varieties of the Ash have recently been

introduced from Japan, but they have not as yet been sufficiently tested to determine their value, but some of them are known to be quite tender in our Northern States.

GLEDITSCHIA, L.—*Honey Locust.*

A genus of handsome deciduous trees, with light and airy foliage. The flowers small and inconspicuous, but are succeeded by one or many-seeded linear, and often twisted pods, containing a sweetish pulp, hence the common name. Our indigenous species are,

Gleditschia triacanthos, Linn.—Three-thorned Acacia or Honey Locust.—Leaflets lanceolate-oblong, of a light, bright-green color. Thorns mostly compound. flattish at the base, and tapering with branches toward the end, very hardy and strong, and on old trees these thorns are often nearly or quite a foot long. Pods ten to twenty inches long, and an inch or more wide, usually slightly twisted. Seeds compressed, very hard and horn-like. The pods often hang on the trees nearly all winter. Seeds do not germinate readily if allowed to become dry, but by scalding and soaking in tepid water for a few days, seed two or three years old can usually be made to grow. The thorns appear on all parts of the tree, and very large ones protrude from the main stem, and larger branches, and these, when they fall off, become dangerous to animals and persons frequenting the ground where the trees are growing; and this is one of the greatest objections to this handsome tree. It is often used for hedges, but when pruned, the twigs and branches should be carefully gathered up and burned, or otherwise destroyed. It is a very large, handsome, clean tree, seldom attacked by insects, and quite hardy in our most Northern States. The wood is heavy, hard, and rather coarse-grained and valuable for many purposes. A variety of this known as *inermis* or thornless, frequently appears among seedlings, and occasionally very large specimens are seen, and are much more desirable than those bearing thorns, but they can only be increased by budding or grafting, as they do not come true from seed. Var. *Bujoti pendula* (Bujot's Weeping) is an elegant, small tree with drooping foliage, coming into leaf quite late in spring. Said to be not quite hardy at Rochester, N. Y., but it is in my grounds, as I have one tree twenty years old, that has never been injured, even in the coldest winters.

The Honey Locust is supposed not to be indigenous east of
the Alleghany Mountains, although very large old trees are
found in nearly all of our Eastern States, probably raised from
seed brought from Western localities. More or less common
from Pennsylvania southward to Florida, and westward to the
Valley of the Mississippi.

G. monosperma, Nutt.—Water Locust.—Leaflets ovate or oblong, thorns mostly simple, not branched. Pods short, oval,
one-seeded, without pulp. A small tree in swamps and low
grounds, from Southern Illinois to Florida. It is occasionally
planted for ornament.

FOREIGN SPECIES AND VARIETIES.

G. Caspica.—Caspian Honey Locust.—A rapid growing tree
of irregular form, but with large foliage. Thornless.

G. Sinensis.—Chinese Honey Locust.—Has stouter and conical
thorns, and broader and more oval leaflets. A small tree, quite
hardy in our Northern States. There is also a thornless variety of this species also hardy.

GORDONIA, Ellis.—*Loblolly Bay.*

Elegant, small shrubs, or large trees, with showy flowers,
closely related to the common *camellia.* Flowers with five
thick petals, imbricated in the bud. Fruit woody, five-valved,
containing rather long, angular, or winged seeds. Mostly
native of the West Indies, one species in the Island of Java,
and two in our Southern States. Propagated from seed, or by
cuttings or layers.

Gordonia Laslanthus, L.—Loblolly Bay.—Leaves obovate-oblong,
narrowed in a petiole, finely serrate ; evergreen. Flowers
silky, two inches broad, white with long stalks, appearing in
July and August. A large tree thirty to sixty feet high, with
a stem nearly two feet in diameter. Wood of a reddish color,
rather light and brittle, not considered valuable. In swamps
of Southern Virginia, Florida, and west to Louisiana.

G. pubescens, L. Herit.—Leaves obovate-oblong, sharply serrate,
white beneath, deciduous. Flowers silky, on short stalks,
white, fragrant, and nearly three inches broad. A small tree
about thirty feet high in Georgia, and near the coast southward. Hardy as far north as Philadelphia, and quite a large
specimen was formerly growing in the old Bartram Garden,
Michaux ; also Meehan, in Hand Book of Ornamental Trees.

GUAICUM.—*Lignum Vitæ.*

A genus of only a few species of evergreen trees and shrubs, of the West Indies, and South America. One of the species (*G. officinale*) yields the peculiar substance known as *guaicum*, used as a medicine. Wood exceedingly hard and heavy. Plants sometimes cultivated in green-houses as a curiosity, and propagated by cuttings which grow quite readily if taken off at a joint and set in sand where they can be given bottom heat. We have one species

Guaicum sanctum, L.—Lignum Vitæ.—Leaflets six to eight, obovate or oblong, tipped with a short point, entire. Flowers clustered at the forks of the branches, about half an inch broad, blue or purple. A small tree in Southern Florida, and the West Indies.

For *G. angustifolium* of Engelmann, see *Porliera.*

GYMNOCLADUS, Lam.—*Coffee Tree.*

A genus of only one species, and its name is from two Greek words, meaning naked branches, for when the large leaves fall in autumn the tree appears destitute of any fine twigs or spray.

Gymnocladus Canadensis.—Kentucky Coffee Tree.—Leaves very large, two to three feet long, twice pinnate, each partial leaf stalk bearing seven to thirteen ovate leaflets, except the lowest pair, which are single and two to three inches long, the leaflets usually hanging edgewise. The color of the leaves is a bluish-green. The flowers are whitish, borne in short spikes, and the two sexes on separate trees, consequently, both must be present in order to secure fruit on one. Fruit an oblong pod, six to ten inches long, one to two inches broad, containing several large, slightly flattened, smooth, hard, nut-like seeds. Wood of a reddish color, compact, very tough, and susceptible of a high polish, but so cross-grained that it can scarcely be split, and when sawed into planks it warps very much in seasoning. A very large tree, with a rough bark, sixty to eighty feet high, and stem two feet or more in diameter. From Western New York to Nebraska, and southward to Tennessee, reaching its greatest size in the latter State and Kentucky. Readily propagated by seeds or root cuttings. A rapid growing tree in moist, rich soils, but succeeds poorly in light, dry soils. Not especially valuable except as an ornamental tree.

HALESIA, Ellis.—*Silver-bell Tree.*

A genus of deciduous shrubs, or small trees, with large, veiny and pointed leaves, and showy white, or pinkish flowers, on long, slender stalks in clusters or short racemes, from axillary buds of the preceding year. Fruit large, and dry, with two to four wings, the shell within very hard and horn-like. Seeds cylindical, and oval in each cell. Three species, and all indigenous to the United States.

Halesia diptera, L.—Two-winged Silver-bell Tree.—Leaves oval, coarsely serrate, four to five inches long, soft, pubescent. Flowers white, about an inch long, on slender pedicels. Fruit compressed, an inch long, with two wings. A small tree or large shrub, with very hard wood. In rich woods of Georgia and Florida, and westward. Not quite hardy in the latitude of New York, but sometimes escapes injury if planted in a protected situation.

H. tetraptera, L.—Silver-bell or Snow-drop Tree.—Leaves oblong, finely serrate, two to four inches long. Flowers two to four in a cluster, nearly an inch long, pure white. Fruit with four wings. A very handsome, small tree, if kept properly pruned, otherwise it will form a large clump with several stems springing from the same root. If kept to a single stem, it will grow thirty or more feet high, with a stem a foot in diameter. Wood light-colored, exceedingly hard, and fine grained. A handsome, ornamental plant, hardy in most of our Northern States. Native of Southern Illinois, Arkansas, and southward to Louisiana, and eastward to North Carolina and Florida.

H. parviflora, Michx.—Small Flowered Snow-drop Tree.— Leaves ovate-oblong, pointed, soft and velvety while young. Flowers four to five in a clustered somewhat leafy raceme. Smaller than the last, and more or less tinged with red or pale-rose. Fruit slender and unequally winged. Michaux gives Florida as its native habitat. It appears to be a rather rare shrub in cultivation, and seldom mentioned in nurserymen's catalogues. I have a specimen plant in my grounds set out twenty years ago, and it has never failed to bloom, showing that this species is quite hardy even in our Northern States. It is merely a large shrub, six to ten feet high, and the stems not as large or stocky as in the other two species.

HETEROMELES,. Rœmer.—*Toyon or Tollon.*

A genus of single evergreen species, closely related to the Hawthorn, but differing in form of the calyx, number and position of the stamens, and other parts of both flowers and fruit. It has been described under seven different specific names, by as many botanical authorities.

Heteromeles arbutifolia, Rœmer.—Arbutus-leaved Toyon.— Leaves deep-green above, light-green beneath, oblong-lanceolate, two to four inches long, on short stalks, margin slightly revolute. Flowers somewhat soft and velvety, with five white, spreading petals · in terminal corymbose panicles. Fruit about a quarter of an inch in diameter, red, and of the same flavor as some of the Hawthorns. A small tree or large shrub, in the Coast Ranges of California.

HIPPOMANE, Linn.—*Manchineel.*

Tropical evergreen trees, abounding in a white, milky juice, which is very poisonous, and if it touches the skin will cause severe irritation and blisters. There are two species, one of which is found on our southern border.

Hippomane Mancinella, L.—Manchineel.—Leaves alternate, ovate-serrulate, pointed, nearly smooth. When the leaves fall off with age, they leave large scars, giving to the branches and stems a very rough appearance. Flowers minute, greenish, without petals, and in short spikes of about two inches in length. Fruit large, somewhat resembling apples, but said to be poisonous. The old botanist, Jacquin, said that they were eaten by the sea crabs. Wood heavy, clear-grained and beautifully variegated, and veined with various shades of brown, white, and yellow. Highly valued for fancy boxes, and other kinds of cabinet work, as the polished surface of the wood resembles some of the finest varieties of marble. In cutting the tree, the workmen have to be very cautious to prevent the juice getting upon their flesh. This tree grows to an immense size in the West Indies, but only forty or fifty feet high in Southern Florida.

HYPELATE.—*Honey-berry.*

A small genus of sub-tropical evergreen trees, principally in the West Indies, where they are cultivated for their small, sweet, and edible berries. One native of Ceylon, and two scatteringly in Southern Florida.

Hypelate paniculata.—Maderia Wood, Honey-berry, Genip-tree.
—Leaves abruptly pinnate, leaflets oblong, entire, two to three
inches long, smooth, deep-green above, and pale beneath.
Fruit round, of a green color, pulp sub-acid, astringent or sweetish when fully ripe. This is the *Melicocca paniculata* of Jussieu.
A small tree in Southern Florida, and the West Indies. Wood
hard and flexible, used for bows and spears.

II. trifoliata.—Three-leaved Genip-tree.—Leaves trifoliate;
leaflets obovate, rather thick, and of a leathery appearance on
the upper surface. Flowers small, white, in a short panicle or
cluster, only a few in number. Fruit black and only one-
seeded. A small tree with brittle branches. Southern Florida
and the West Indies.

ILEX, Linn.—*Holly.*

A very extensive genus of small trees and shrubs, mostly with
thick and rigid evergreen leaves, small white flowers and red
berry-like fruit. Hollies are to be found in nearly all parts of
the world, but 'mainly in temperate climates, but most highly
prized as ornamental plants in Great Britain, where, from the
European Holly (*Ilex Aquifolium*), scores of elegant varieties
have been produced and extensively propagated. These European
varieties are not well adapted to our dry and hot climates,
and are seldom cultivated, except in conservatories, or in such
positions where they can be protected in winter, and shaded
from the hot sun in summer. We have only two species that
grow to the size of trees.

Ilex Dahoon, Nutt.—Dahoon Holly.—Leaves acute or obtuse,
serrate, or toothed with sharp points, young branches and lower
surface of the leaves more or less pubescent. There are several
well-marked wild varieties. A small tree, with very hard
wood, sometimes twenty-five or thirty feet high. Virginia to
Florida and westward.

I. opaca, Aiton.—American Holly.—Leaves oval, concave,
wavy, and sharp spines on the margins. Flowers at the base
of the previous season's shoots, succeeded by bright red berries,
which remain on the tree all winter, and are much sought
after about the holidays for decorating churches and private
dwellings. A tree from twenty to forty feet high, wood very
hard. Found sparingly in Southern New England, Long Island,
New Jersey, and southward to Florida. One of our most
beautiful broad-leaved evergreen trees, scarcely hardy north of

the latitude of New York City, except in protected situations. I have specimens in my own grounds standing on the south side of Arbor Vitæs and other positions, where they are shielded from north-west winds, that have never been injured, but others not a hundred feet distant, are often badly browned in winter. Our American Holly is well worthy of more extended cultivation than it has ever received, and should be planted as an under-shrub in forests, wherever the climate will permit. The seed should be stored in moist earth or sand for one year before sowing, for if sown as soon as gathered, they will not sprout until the second year. The Hollies are readily propagated by grafting in spring, or budding in the latter part of summer. The weak-growing or dwarf species and varieties being worked on the strong.

I. Cassine, L.—Yaupon Holly.—Leaves small, one-half to an inch long, oval or oblong, with the edges scalloped into rounded teeth. Fruit very abundant, and in clusters. The leaves of this species are sometimes used as a substitute for genuine tea, and are known in the South as Yaupon tea. A small shrub, only eight to ten feet high, in sandy soils from North Carolina southward.

I. glabra, Gray.—Common Inkberry.—Leaves smooth, wedge oblong or obovate, slightly toothed near the apex. Fruit small, black. A small shrub along the coast of New England, New Jersey, and southward to Florida.

I. coriacea, Prinos coriacea, Ell.—Leaves somewhat like the last, but with sharp scattered teeth on the margin. A small shrub in wet places in the South.

We also have some four or five deciduous species of the *Ilex* or *Prinos* of some botanists, but all are shrubs, mostly frequenting swamps and low grounds. The *I. verticillata*, Gray, is known as Black Alder or Winterberry, the bright scarlet berries in autumn and early winter make this shrub a very conspicuous object in swamps and low grounds in our Northern as well as Southern States.

JUGLANS, Linn.—*Butternuts, Walnuts.*

In species this genus is quite restricted, there being but one indigenous to the eastern hemisphere, and four to the western, and all these natives of the United States. They are large trees with one exception, bearing edible nuts, some of which are highly prized and extensively cultivated. Flowers similar to

those of the Hickory (*carya*), the staminate catkins produced
from the previous year's wood, long, solitary, or in pairs. Fertile flowers solitary, or few in a short terminal spike; the calyx
adhering to the ovary. Fruit fleshy, enclosing an irregularly
rough nut. Wood of all valuable.

Juglans Californica, Watson.—California Walnut.—Leaves more
or less downy. Leaflets five to eight pairs, oblong-lanceolate,
acute, narrowing upward from near the base, and two to two-
and a half inches long. Fruit round, slightly compressed,
about an inch in diameter, shell rather thin, with two broad
cavities upon each side. A tree or large shrub near San Francisco, and on the Sacramento a tree forty to sixty feet high, and
stems two to four feet in diameter. Also in Southern California, Arizona, New Mexico, and in Sonora, Mexico.

J. cinerea, L.—Butternut.—Leaflets oblong-lanceolate, pointed,
rounded at the base, downy, especially underneath, and the
petioles and branchlets with clammy hairs. Fruit oblong,
clammy, and the nut deeply sculptured and with ragged sharp
ridges ; kernel sweet, rich, and oily. A well-known tree with
gray bark, and only slightly furrowed on the stems of old
trees. Wood light-colored, only moderately hard, very durable, and considered valuable for cabinet work and various
other purposes. The inner bark has long been used for coloring cloth, and the historic "Butternut color" is not quite ex-
tinct, although not so common as it was a half century ago.

A large tree in the bottoms along our northern rivers ; sometimes sixty feet in hight, and stem two feet or more in diameter. A rapid growing tree, readily raised from the nuts, and
can be safely transplanted at almost any age, especially when
raised in nurseries and moved when young. A common tree
in nearly all of our Northern States, and southward along the
mountains.

J. nigra.—Black Walnut.—Leaflets eleven to twenty-one, ovate-
lanceolate, slightly pubescent beneath, pointed, slightly heart-
shaped at base ; neither leaves, stalks,. or fruit clammy, as in
the last. Fruit large, round, somewhat dotted, but not furrowed. Shell of nut black, or dark brown, very rough ; kernel
large, very oily, and a strong, rather disagreeable flavor, but
not at all poisonous as sometimes stated. Wood of a dark,
rich brown color, rather hard and firm, but susceptible of a
high polish, and probably more extensively employed for first-
class cabinet work than any other native wood. It is also ex-

tensively used for gun-stocks, hand-rails, floors, stairs, and inside finishing generally. One of our most valuable timber trees, once so abundant in some of our Western States, as to be employed for fencing, and many farms have been enclosed with fences made of black walnut rails. If the trees had been left standing, they would now be worth many times more than the land is, from which they were so ruthlessly destroyed. The Black Walnut is a noble tree with a very erect straight stem, often reaching a hight of sixty to ninety feet, and from four to six feet in diameter. The bark is usually rough, dark-colored and deeply furrowed. If raised in nursery rows, and the seedlings transplanted, and roots pruned when young, the Black Walnut may be moved without danger of loss, when from four to six feet high. For planting, the nuts should be gathered in the fall, mixed with soil or sand, and left in heaps exposed to frosts during the winter. In spring plant them in rows, covering them with an inch or two of soil. The Black Walnut is a rapid-growing, hardy tree, commencing to bear nuts in eight to ten years, but will require from twenty to forty years to reach a size large enough to produce boards or planks. It is widely distributed over the United States, from Vermont to Florida, and westward to Texas, thence northward to Nebraska, but now more abundant west of the Alleghanies than east of them.

J. rupestris, Engelmann.—Leaves composed of from six to twelve pairs of leaflets, usually short-pointed. Nut very small, round, very thick, nearly solid walls. A small tree or shrub, ten to twenty feet high, in Western Texas, New Mexico, and Arizona.

FOREIGN SPECIES AND VARIETIES.

Juglans regia.—English Walnut, French Walnut, Madeira Nut.—Leaves composed of from five to nine oval, smooth leaflets. Fruit round or oval, and when ripe, the husk becomes friable and brittle, opening and allowing the nut to fall out. The shell is thin, kernel large, with a rich, oily, but rather strong taste. A well-known nut in our markets, and throughout those of the greater part of the whole world. Although often called English or French Walnut, the tree is not a native of these countries, but is found in Persia, and according to some authorities in China. It has been cultivated for so many centuries in Europe, that a large number of varieties have been produced there, varying much in the size and quality of the nuts, also in the leaves, and form and growth of the trees. In general appearance this Royal Walnut of the East resembles our American Butternut, but the trees grow to even a larger size. The wood is quite valuable and highly prized in Europe for cabinet work and similar uses. But after the discovery of America and the introduction from here of Black Walnut, which is its superior for similar purposes, its value decreased in consequence of competition, but even now is much sought after and commands a large price. The success of this tree in the United States has been greatest in the Middle and Southern States, still there are many quite old and large trees in the suburbs of New York City, some of which seldom fail to produce a good crop of nuts. In years past there has been no especial attention given to the selection of varieties adapted to the climate of our Northern States, hence there has been more

Fig. 38.—SMALL FRUITED WAL-NUT.

Fig. 39.—GIBBOU'S WALNUT.   Fig. 40.—BARTHERE WALNUT.

failures than successes in attempts to cultivate this Eastern nut, but with proper care in selecting seedlings from the cooler parts of Europe, and propagating from well-tested varieties that do succeed here, I can see no good reason why this nut should not be raised in abundance in most of our Northern States. Recently a dwarf French variety, called the "Preparturiens," has been introduced and largely propagated by our nurserymen, as it is said to come into bearing when quite young. There is a very large number of varieties cultivated in the French and other European nurseries, the nuts varying greatly in size and form, as shown in the accompanying illustrations, which I have selected merely to give the reader an idea of the extremes of variations to be found among the nuts. The Small-fruited Walnut (*Juglans regia microcarpa*), shown in fig. 38, is only about one-half inch in diameter, but the kernel is sweet, and of a most delicate flavor and texture. This is probably the smallest variety known, while the Gibbou's

Fig. 41.—CUT-LEAVED WALNUT.

Walnut (*J. regia Gibbosa*), fig. 39, is quite the opposite or the largest, and while it has a rather thick shell, the kernel is of excellent quality and easily extracted. The Barthere Walnut (*J. regia Barthereana*), fig. 40, is remarkable for its extreme length. Its shell is quite thin, and the flavor of the kernel is excellent. Between these extremes of variation there are scores of others that are quite distinct, and may be found described in the catalogues of European nurserymen. There is a also a marked difference in the habit of growth of the different varieties, some being very tall trees, and others dwarfish. There are also broad-leaved and narrow-leaved varieties, and among the latter the pretty Cut-leaved Walnut (*J. r. laciniata*), fig. 41.

KALMIA.—*American Laurel.*

A genus of less than a half dozen species, all native of North America. Two are low shrubs, and only one growing to the hight of twenty feet or more. They all have handsome evergreen leaves and showy flowers. The common name "laurel" is a misnomer, as the Kalmias are far removed botanically from the true Laurels (*Laurus*). The genus was named for Peter Kalm, a pupil of Linnæus, who visited this country early in the last century.

Kalmia latifolia, L.—Broad-leaved Kalmia, Calico-bush, Spoonwood.—Leaves oval or lance-oval, bright green on both sides. Flowers in large, showy clusters ; white to deep pink or rose, with crimson spots, appearing in spring or early summer. One of our most beautiful native shrubs or small trees. Common in damp soils and rocky woods, from Maine to Florida. Usually a low spreading shrub, but in favorable soils in the Middle States, it reaches a hight of thirty feet, with a stem a foot or more in diameter. Wood very hard and close-grained ; excellent for handles of tools, wooden spoons, etc.

K. angustifolia, L.—Narrow-leaved Kalmia or Sheep Laurel.— Leaves very narrow, with short stalks, pale-green beneath. Flowers smaller than the last, of a crimson purple color, appearing late in spring. Has the reputation, probably unjustly so, of poisoning sheep and calves, and for this reason has received the common names of Lamb-kill and Kill-calf. A shrub from two to three feet high, in low, dry grounds. New England to Alabama.

K. cuneata, Michx.—Carolina Sheep Laurel.—Leaves sessile, alternate, wedge-shaped, pubescent beneath, bristle-pointed. Flowers white. This is a rare shrub and may be only a variety of the last. Swamps of North and South Carolina.

K. glauca, Ait.—Pale Laurel.—Leaves opposite, oblong, whitish beneath, with revolute margins. Young branches two-edged. Flowers in small terminal clusters, of a lilac-purple color. A small shrub, only one or two feet high in cold bogs, from Pennsylvania northward.

K. hirsuta, Walt.—Hairy-leaved, Wicky.—Leaves alternate opposite, hairy, and only a half inch long, oval or oblong. Flowers numerous but solitary, and of a pale rose-color. A shrub six to eighteen inches high, in pine-barrens. Georgia and Florida.

LAGUNCULARIA, Gært.—*White Mangrove.*

Small maritime, sub-tropical trees or shrubs, with evergreen, opposite leaves, of an elliptical form, and thick and fleshy. Flowers small, yellowish white, in simple or compound axillary and terminal spikes. Fruit a drupe, with one seed or nut. We have but one, or at most, two species.

Laguncularia racemosa, Gært.—White Mangrove, Black Button Tree.—Spikes upright, rigid, hairy, the lateral ones solitary, the terminal ones in threes, simple or branched. Flowers scattered. A small tree or shrub in South Florida and the West Indies. The *L. glabriflora* of Presl. is probably only a form of the above, found in the same regions, neither of any value to man so far as known.

LIQUIDAMBAR, L.—*Sweet Gum.*

Deciduous trees, with monœcious flowers, in globular, four-bracted spiked heads. The flowers are very small, and have neither calyx or corolla, but sterile ones with numerous stamens. Heads of sterile flowers sessile, crowded, those of fertile ones on a long drooping peduncle. Seeds small, angled or scale-like. One species belonging to this country.

Liquidambar Styraciflua.—Sweet Gum, Bilsted, Alligator-tree.—Leaves roundish, but with five to seven-pointed spreading lobes. In autumn they assume a rich bronze color, but on some trees they change to a crimson. The smaller branches are ornamented with prominent corky ridges, and the young twigs can often be selected of very curious shapes, having a fanciful resemblance to some of our reptiles, and this may have suggested the name of Alligator tree, under which name the twigs are frequently sold in the streets of New York. The Sweet Gum is one of our most noble forest trees, somewhat resembling the Sugar Maple, but with a more conical head, the branches spreading widely, often drooping, with the ends curved upward. It is also a rapid growing tree, and thrives on a great variety of soils, from the light, dry, and sandy, to the cold and wet. Among the first trees planted on my lawn was one Liquidambar, and I have never regretted giving it a conspicuous position, as it is one of the very best ornamental trees in my collection. The wood of this tree is very light, but compact, fine grained, but not hard, sometimes used for cabinet work, but owing to its softness is easily bruised. It is what is termed uneven-grained

wood, warping badly when sawn into boards or planks. It decays rapidly when exposed to the weather, and is of very little
value for fuel. Although this is not a valuable timber tree, it
is well worth cultivating for ornament. A large tree, sixty to
eighty feet high, with a very straight stem. More abundant in
the Eastern and Southern States than in the West. Native of
New England and southward to Central America. Varieties of
this species appear to be quite rare, and I do not now call to
mind any that have been disseminated by our nurserymen.
The late Joseph Longworth, of Cincinnati, Ohio, found a curi-

Fig. 42.—LONGWORTH'S LIQUIDAMBAR.

ous variety several years ago growing in the suburbs of that
city, and transplanted it to his grounds. The peculiarity of
this variety was in the form of the leaves, as shown in fig. 42,
and instead of being five-pointed, star-shaped as usual in the
species, they have only three prominent points with one or two
smaller ones. I think that this variety was first described by
the editor of the *American Agriculturist* in 1868, who also suggested that it should bear the name of *L. Longworthii.*

There is an oriental species of the Liquidambar (*L. orientalis* Mill, or *L. imberbe*, Aiton), and although introduced into the gardens of England in 1759, it has never become popular as an ornamental plant, probably because of its dwarfish habit, seldom growing more than a dozen feet high.

LIRIODENDRON.—*Tulip Tree.*

A magnificent native deciduous tree, belonging to the Magnolia family. Flowers composed of six petals, bell-shaped. Fruit a cone-like head, comprised of a large number of dry seeds, with long, narrow scales or wings, attached to a common axis at their base, forming a conical spike two inches or more in length. Each fruit is composed of sixty or more winged seeds, but only a small proportion contain kernels, or will germinate. Seeds from old trees are less productive than those from young ones. There is but one species.

Liriodendron Tullpifera.—Tulip Tree, White-wood, Canoe-wood, Virginia Poplar.—Leaves large, smooth on both sides, on slender stalks, somewhat three-lobed, the middle one appearing as if it had been cut off, leaving a shallow notch. Flowers bell-shaped, greenish-yellow, tinged with orange. The branches with smooth, grayish bark. Terminal buds on the shoots swell considerably in spring, before the leaves unfold. On young thrifty trees the leaves are often six to eight inches in diameter, and of a clear, bright-green color. Bark on old trees deeply furrowed, and quite thick. One of our finest and largest forest trees in the Eastern States, sometimes reaching a hight of one hundred and forty feet, with a stem four to six feet in diameter. Wood light, soft, but close-grained, easily worked, and extensively used for the interior work of carriages, furniture, and other purposes where it is not exposed to air and moisture. It is only valuable for inside work, for it decays rapidly if exposed. This is one of the few kinds of wood that will shrink endways of the grain when seasoning. The wood is also quite variable in color and texture, and that from the largest trees that have grown on rich soils is the most valuable. The tulip is readily propagated from seed, and if several times transplanted in the nursery, the trees produce a great abundance of fibrous roots. Thrives best on light, deep loam, or sandy soils. Does not succeed in a heavy clay or in swampy land. It is what may be called a very clean tree, only a very few species of insects attacking it, and these appearing to do it but little injury. Native of Vermont and southward to Florida, and westward to Eastern Kansas. It was formerly quite abundant in Western New York, and where very large trees were plentiful, but is now scarce, except in some of the forests of the Middle States.

MACLURA, Nutt.—*Osage Orange.*

A genus of handsome ornamental trees of moderate size, with diœcious flowers, the staminate or male in long racemes, resembling those of the common Mulberry, the pistillate or fertile flowers densely crowded in a large, spherical bead, becoming a compound globular fruit, resembling a large, rough orange of a greenish yellow color, containing a large number of obovate seeds. Two evergreen species in the West Indies and one deciduous in the United States.

Maclura aurantiaca, Nutt.—Osage Orange, Bois d'arc.—Leaves alternate, entire, rather long pointed, bright, glossy green, usually with a sharp spine at the base of each, even on the smallest twigs. Wood solid, heavy and elastic, quite durable, of a fine yellow color, which is readily communicated to water. Usually a medium-sized tree, but sometimes grows fifty feet high, with a stem two feet in diameter. Abundant in the South-western States, Eastern Texas, Arkansas, Indian Territory. This tree has been more extensively employed for hedges than any other, and is well adapted for live fences, as the thorns, while abundant, are not so large and strong as to be dangerous to stock. It is not quite hardy in our more Northern States, although in sheltered positions the trees grow rapidly, and to a considerable size, fruiting quite freely in the neighborhood of New York City. The introduction of barbed wire fencing has made hedge plants less a necessity than formerly, and the Osage Orange will probably not be very extensively cultivated in the future, except as an ornamental tree, or for feeding the silk-worm, for which the leaves are well adapted

MAGNOLIA, Linn.

An extensive and widely distributed genus of trees and shrubs, with large and showy flowers. There are both deciduous and evergreen species, and the larger proportion, and those growing to the largest size, are natives of the United States. They are usually propagated by seeds or layers, although both budding and grafting are practiced in multiplying rare species and varieties. The seeds should never be allowed to get thoroughly dry, but as soon as removed from the pulp, be mixed with moist sand or soil, or sown immediately. Layers put down in spring root freely, but it is well to allow the layered branch to grow undisturbed through the entire first season, neither cutting it back or removing any of the leaves. Budding may be done in summer while there is a rapid flow of sap, but grafting is generally most successful upon stocks grown or kept in a propagating house until the cion has united, and made one season's growth. In grafting the evergreen species, the cion should be inserted in the side of the stock and below some good healthy leaves, somewhat after the manner of grafting conifers. Magnolias should never be transplanted in the autumn, especially in cool climates, as their roots are quite soft, and the exposed wood and small fibers decay very quickly on trees transplanted in the fall.

Magnolia acuminata, L.–Cucumber Tree.–Leaves oblong, pointed, green above, but slightly paler beneath, five to ten inches long. Flowers pale-yellowish-green, about three inches broad. Fruit irregular, oblong, containing a few hard, bony, black seeds. A handsome, erect-growing, stately tree, sixty to ninety feet high, with stem two to four feet in diameter. Wood rather soft, of a yellowish-white color, quite durable, and extensively used for pump logs; wooden bowls, and other household utensils are also made from it. Formerly very abundant in Western New York and southward along the mountains to Georgia and Kentucky.

M. cordata, Michx.–Yellow Cucumber Tree.—Leaves oval or roundish, seldom cordate as the name implies, four to six inches long, white, downy beneath. Flowers four to six inches broad, petals six to nine, of a lemon-yellow color. Fruit oval or oblong, about three inches long. A rather broad, spreading tree, thirty to forty feet high, quite hardy as far north as New York, but native of the mountains of North Carolina, and southward to Alabama and Georgia. It is not a very popular ornamental tree, and the wood is too soft and light to be of much value.

M. Fraseri, Walt.—Ear-leaved Magnolia.—Leaves nearly a foot long, spatulate-obovate, smooth on both sides, heart-shaped, and two-eared or auricled at the base; stalks slender. Flowers about six inches broad, white and fragrant. A tall, rather slender tree, quite hardy, often blooming twice in a season in the vicinity of New York. Wood soft, but resembling that of the first.

M. glauca, L.—Sweet Bay, Swamp Magnolia.—Leaves quite thick, oblong-oval, smooth and glossy above, white or rusty pubescent beneath, evergreen in the South, and nearly so in protected situations at the North. Flowers composed of nine concave petals about two inches broad, white, and very fragrant. Cone of fruit oval, about an inch and a half long, containing numerous black seeds, enclosed in a light scarlet pulp or aril, a character common to most of the species, but in some the aril is of a darker color. The Sweet Bay or Swamp Magnolia is one of our most beautiful ornamental shrubs, or small trees, and while it thrives best in low, most soils, it will grow quite well in any moderately good garden soil. A small tree, but often twenty to thirty feet high in swampy grounds. A variety of this, known as the *M. longifolia*, has larger leaves than the species, otherwise not different. Native of Massachusetts, and southward to Florida.

M. grandiflora, L.—Large-flowered Magnolia, Southern Evergreen Magnolia.—Leaves evergreen, thick and leathery, oblong, smooth above, rusty pubescent beneath, six to twelve inches long. Flowers white, fragrant, and from six to ten inches broad. Fruit oval, three to four inches long. Wood soft, and very white, of little value except for inside work, and where it will not be subjected to any wear. A large tree from fifty to nearly a hundred feet high, with stem two to three feet in diameter. This is, without doubt, the most noble, broad-leaved, evergreen tree found in North America. Unfortunately it is not hardy in our Northern States, and must be treated as a green-house shrub, or at least given some protection in winter. Native of North Carolina, and south to Florida, thence west to Texas, and in the Mississippi Valley as far north as Natchez.

M. macrophylla, Michx.—Great-leaved Magnolia.—Leaves very large, sometimes three feet long, usually clustered on the ends of the stout, cane-like, whitish, pubescent branches. The leaves are broadest above the middle, or obovate-spatulate, heart-shaped, or slightly eared at the base, green above, but whitish beneath. Flowers white, with a purple spot near the base, fragrant, and often twelve inches broad. Fruit nearly cylindrical, and about four inches long, the color at maturity is

8

a deep rose. A tropical-looking tree, growing thirty to forty feet high. Wood soft, and of little value. A rare tree in nature, and not found anywhere in great abundance, but scatteringly in North Carolina to Florida, and Kentucky and Tennessee. A tree thirty to forty feet high, with a stem a foot or a little more in diameter. Quite hardy in the vicinity of New York City, but tender farther north.

M. Umbrella, Lam., *M. tripeleta,* L.—Umbrella Tree.—Leaves clustered at the ends of the branches, obovate-oblong, twelve to eighteen inches long, pointed; downy beneath, but becoming smooth with age. Flowers of about nine petals, white, six to eight inches broad. Fruit oblong, four to six inches long, rose-colored when mature; quite ornamental. A small, rather straggling growing tree, thirty to forty feet high. Western New York, Alleghany County in the hills, and southward in the mountains of Pennsylvania, the Carolinas, Northern Alabama and Georgia, and westward to Kentucky and Tennessee. A hardy and very handsome tree. Wood rather soft and of little value.

M. Thompsoniana.—Thompson's Magnolia.—Supposed to be a hybrid between *M. glauca and M. Umbrella.* A medium sized tree, with the habits of the last, but blooming irregularly throughout the summer. Flowers large, creamy-white and fragrant. Propagated by grafting on the stocks of *M. acuminata,* which is also the best stock upon which to work nearly all of the species and varieties in cultivation, including the

FOREIGN SPECIES AND VARIETIES.

Of these there are quite a large number, mainly from China and Japan, most of which are hardy in our Northern States. Authorities do not agree as to which should be considered as species or varieties, but as they are all cultivated as ornamental trees or shrubs, and not for any economic purpose, I will only name a few of the best known without regard to their botanical classification. The flowers of the following appear early in spring, before the leaves, or with them, and are very showy when not cut off by frosts, as they often are—in and above the latitude of New York.

M. atropurpurea.—Dark purple flowers, blooming rather late or with the opening of the leaves.

M. conspicua.—Yulan; or, Chinese White.—Flowers very large, white ; appearing very early. Quite a tall shrub or tree when worked on strong growing stocks.

M. Lennei.—Lenne's Hybrid.—Flowers large and showy, crimson outside, and pearl-colored within.

M. Norbertiana.—Norbert's Hybrid.—A seedling of the next, with dark purple flowers.

M. Soulangeana.—Soulange's Hybrid.—A low spreading tree with large whitish flowers, with purple at the base of the petals.

M. speciosa.—Showy Magnolia.—Flowers smaller than the last, but appearing a week later, but similar in color.

M. stricta.—Great Chinese M.—Flowers slightly tinted with purple, but almost white, an erect growing and free-blooming variety.

M. superba.—Superb Chinese M.—Flowers darker than those of *Soulangeana*, otherwise quite similar.

M. hypoleuca.—Japan M.—A recently introduced species, with very large leaves sometimes tinted with purple. A very erect growing tree, with creamy-white fragrant flowers, appearing after the leaves have expanded.

M. Kobus.—Thurber's Japan M.—A medium sized bush with fragrant blush-white flowers.

M. parviflora.—Small-flowered Japan M.—A new variety with small, very fragrant flowers. Leaves large and handsome.

M. purpurea.—Purple Japan M.—An old dwarfish variety, with dark, purple flowers, rather tender. A variety of this known in catalogues under the name of *M. gracilis*, has very slender, upright stems. It blooms freely in my grounds when given a slight protection in winter.

M. stellata.—Star Magnolia, Hall's Japan M.—A low-growing shrub, with pure white fragrant flowers, of a rather loose and irregular shape, but appearing in spring before those of any other species.

MELIA, Linn.—*Pride of India.*

A genus of handsome tropical or sub-tropical trees. Principally evergreen, with large, handsome pinnate or doubly-pinnate leaves. While there are no species native of the United States, there is one that was so early introduced into the Southern States, it has run wild, and become so fully naturalized that some persons suppose it to be indigenous in the Gulf States. This species is

Melia Azedarach.—China Tree ; or, Pride of India.—Leaves very long, doubly-pinnate, dark green, coming out early in spring. Flowers small, but in large axillary clusters, deliciously fragrant. Fruit large as cherries, round, yellow when ripe, eaten with avidity by birds, especially by the robin in its migration southward in the autumn. A handsome, rapid-growing tree, often reaching a hight of forty feet, and a stem eighteen inches in diameter. Wood of a reddish color, resembling some species of the ash, quite durable, and makes excellent fuel. It grows so rapidly that seedlings often reach a hight of ten to fifteen feet in three or four years. It thrives in dry soils, and is planted almost everywhere in the South as a shade tree, and is a universal favorite. Not hardy north of Virginia. A native of Persia, but at what date introduced into this country is not known.

MIMUSOPS, Linn.—*Nasebury.*

A small genus of evergreen trees and shrubs, with milky juice, principally natives of Tropical America, India and New Holland. Fruit of most of the species edible, at least so considered by the people where it is produced. One species indigenous to the United States, the

Mimusops Sieberi, A. DC.—Naseberry.—Leaves rigid, oblong, emarginate at the apex, rather broad or blunt at the base, on stout stems. Flowers small, white, among the clustered leaves on the ends of the branches. Fruit a roundish, many-seeded berry, about the size of a nutmeg ; edible when fully ripe. A small tree at Key West, Florida, but in Jamaica it reaches a hight of forty to fifty feet, and the wood is considered one of the strongest and best in the island.

MORUS, Tour.—*Mulberry.*

A genus of only a few species from which a great number of varieties have originated. Flowers monœcious, the sterile and fertile in separate spikes. Fruit edible, usually oblong, somewhat resembling in structure and form, the common blackberry. Trees or shrubs with milky juice. We have but one, or at most two, native species.

Morus rubra, L.—Red Mulberry.—Leaves broad, heart-shaped, serrate and rough above, and downy underneath. On young shoots the leaves are variously lobed. Fruit dark red, turning to purple when fully ripe. Wood yellow, very heavy and durable, valuable for fence posts, much used when obtainable for tool handles. Usually a small tree, but sometimes found sixty to seventy feet high, with a stem two feet in diameter. Found in no considerable abundance anywhere, but distributed over the country from Western Massachusetts and Vermont, west to the Rocky Mountains, and south to Florida, Texas and Mexico.

M. microphylla, Buckley, is probably only a southern form or variety, with smaller and rougher leaves. Fruit small, sour and black. Texas, and westward to Arizona.

FOREIGN SPECIES AND VARIETIES.

Morus alba, L.—White Mulberry. —Leaves heart - shaped, pointed, serrate, smooth and shining. Fruit white, sweet, but rather insipid. A tree early introduced into the United States, and is naturalized and run wild in the Eastern States. A low growing tree, but with stem from one to three feet in diameter. There are more than a dozen distinct varieties in cultivation. Among the oldest and best known, I may name the *M. multicaulis*, supposed to be one of the best for feeding silk-worms. Rather more tender than the species, the latter being quite hardy in nearly all of our Northern States, while the former is often winter killed, even in the latitude of New York City. The Downing's Everbearing Mulberry is a seedling of the multicaulis, but with very large black fruit, of a rich, sprightly subacid flavor. *M. alba*, var. *tartarica*, has recently been highly extolled as a timber tree, under the name of Russian Mulberry. It is a rapid growing tree, readily propagated by cuttings or seed, and is said to thrive in the dry soils of the western prairies, where it is quite extensively cultivated by the Mennonites, who brought it with them from Russia, but the same tree has long been known in our Eastern States as the Tartarian Mulberry. The mulberries are handsome trees of rapid growth, although they seldom reach a large size. The leaves of the White Mulberry. and many of its varieties, have for ages been used for feeding the silk-worm in China and other countries. The larger-leaved varieties are preferred to the smaller for feeding the worm, and some are more tender and better adapted to this purpose than others. All the species and varieties of the mulberry put out their leaves late in spring. The West India Mulberry, *M. tinctoria*, yields the well known Fustic wood of commerce.

MYRICA, Linn.—*Sweet Gale.*

A genus of small trees or shrubs, mostly evergreen, and the species are pretty widely distributed over the world. Flowers monœcious or diœcious, with both sexes in short, scaly catkins. Leaves usually fragrant, and the fruit a drupe-like nut. Of no special interest to the arboriculturist, further than the tallest species of the genus is a native of the United States.

Myrica Californica, Cham.—Wax Myrtle.—Leaves evergreen, leathery, usually pubescent beneath, oblanceolate, two to four inches long, pointed. Fruit purple, thinly coated with grayish wax, and only about one eighth of an inch in diameter. Native of Northern California to Washington Territory. A tree thirty to forty feet high, with a stem sometimes two feet in diameter. A dwarf species (*M. Hartwegi*, Watson), is found near Sacramento, and in other parts of California, but it is a low shrub. This genus is represented by several small shrubs in our Eastern States, among the best known and most common, are the Bayberry (*M. cerifera*), and the Sweet Gale (*M. Gale*), the latter being also a native of Europe.

MYRSINE, L.—*Florida Myrtle.*

Evergreen trees or shrubs, with mostly entire leaves, and regular monœcious, or diœcious white, or colored flowers. Fruit resembling small plums, commonly with one reddish seed or nut, concave at the base. The species are widely distributed throughout the globe in tropical, or sub-tropical climates, we have one species.

Myrsine Rapanea, Rœm. & Schult.—Florida Myrtle.—Leaves two to three inches long, thick, oblong-ovate, entire, narrowed at the base into a short petiole. Flowers small, white, and in clusters. Fruit less than a quarter of an inch in diameter. A small tree, sometimes twenty or more feet high in the Florida Keys, and through the West Indies to Brazil.

NUTTALLIA, Torr. & Gray.

A genus closely related to the Plum and Cherry, containing only one species. Usually a shrub, with entire deciduous leaves, with white flowers, in loose drooping racemes, which appear with the branchlets from the same buds.

Nuttallia cerasiformis, Torr. & Gray.—Oso Berry.—Leaves broad, oblanceolate, sharp pointed, two to four inches long.

Flowers greenish-white, one fourth of an inch broad. Fruit blue-black, bitter, ripening in June and July. Usually a shrub, but in favorable locations reaching a hight of twenty feet. Coast Ranges of California and northward to Puget Sound.

NYSSA, L.—*Sour Gum.*

A genus of North American deciduous trees, principally in swamps and low, moist soils. Flowers small, greenish, sterile ones numerous in clusters, the fertile, solitary or few in a bud. Fruit a one-seeded drupe, in some species edible. Usually propagated by seed or layers, but the wild plants can be obtained in abundance.

Nyssa capitata, Walt.—Ogeechee Lime.—Leaves three to five inches long, oblong on short petioles, whitish beneath. Flowers below the leaves, the fertile ones solitary, on short stalks. Fruit about an inch long, oval, red, and the pulp of an agreeable, sub-acid flavor. The conserve known as the "Ogeechee Lime," is prepared from this fruit. Swamps of Georgia and Florida, and westward. A tree thirty feet high, with very tough, cross-grained wood.

N. Caroliniana, Poir.—Carolina Gum Tree.—Leaves from one to two inches long, broad, lanceolate, sometimes slightly heart-shaped at base. Fruit small, dark-blue. A large tree in Southern swamps, with moderately firm, close-grained wood, very difficult to split, and for this reason is much used for hubs and similar purposes. The leaves turn to a brilliant crimson color in autumn, making these trees very conspicuous objects in the forests at that season. North Carolina to Florida in swamps and low grounds.

N. multiflora, Wang.—Tupelo, Pepperidge.—Leaves oval, rather thick, and dark-green, two to five inches long. Sterile flowers in loose clusters, fertile clusters long and slender; containing from three to eight flowers. Fruit ovoid, dark-blue, about a half an inch long. A large tree fifty to sixty feet high, with stem two feet in diameter. Wood tough, cross-grained, difficult to split, used for hubs and similar purposes. A handsome ornamental tree, growing rapidly in moist soils, the branches spreading widely at right angles from the stem. Leaves change to a bright crimson in autumn. Common in low grounds from Vermont, New York, New Jersey, and southward to Florida, also in the Western States.

N. sylvatica, Marsh.—Black Gum.—This is a doubtful species described by several botanists under different names, but Watson (in Botanical Index) considers it only a variety of the last. Michaux found it growing near Philadelphia, and further South on rather high and dry grounds, among oak and walnut trees. Leaves five or six inches long, alternate, oblong-oval. Fruit deep-blue. Wood fine-grained, but rather soft, very cross-grained. A large tree sixty to seventy feet high, with stem two feet in diameter.

N. uniflora, Wang.—Large Tupelo, Cotton Gum. — Leaves large, four to six inches long, ovate or oblong, sharp pointed, entire or sharply toothed, downy beneath. Fertile flowers solitary. Fruit ovate-oblong, dark-blue. A large tree in swamps, from Virginia southward. Wood cross-grained, light, and of little value.

OLNEYA, Gray.—*Iron-Wood.*

A small tree belonging to the Leguminosæ, bearing bean-like fruit, and pinnate leaves, resembling those of the Locust. Only one species.

Olneya Tesota, Gray.—Iron Wood.—Leaves composed of from five to seven pairs of wedge-shaped, oblong leaflets. Flowers pea-shaped, white or purplish, three or four in a loose raceme. Fruit a rough, linear, oblong pod, about two inches long, containing one or two ovate seeds. This is the *Arbol de hierro* or Iron Wood of Arizona, and adjacent regions in California.

OSMANTHUS, Benth. & Hook. OLEA, Linn.

A tree closely allied to the Olive (*Olea Europœa*), and usually called the American Olive. Trees or shrubs with mostly entire leaves, and perfect flowers, but in some they are diœcious, and usually small, white, in cluster or panicles. One native species.

Osmanthus Americanus.—Devil-wood.—Leaves oblong, lanceolate, smooth and shining, three to six inches long. Flowers in compound racemes, shorter than the leaves. Flowers small, white, and fragrant. Fruit ovoid, dark-purple, about the size of a pea, bitter and astringent. A small tree, with very hard, iron-like wood, seldom over twenty feet high. In moist woods from Southern Virginia to Florida.

OSTRYA, Micheli.—*Hop-Hornbeam.*

A tree closely related to the common Beech tree, but with
the fertile flowers numerous, in short terminal catkins, with
small deciduous bracts, each enclosed in a sac-like involucre,
which enlarges and forms an imbricated strobile, like that of
the common Hop. Slender tree with very hard wood.

Ostrya Virginica, Willd.—American Hop-Hornbeam, Iron-wood,
Lever-wood.—Leaves oblong-ovate, taper pointed, very sharply
and doubly serrate, downy beneath. Flowers minute, appearing with the leaves. Seeds in short imbricated catkins, as
shown in figure 43, which are about one half the natural size. A

Fig. 43.—HOP-HORNBEAM.

handsome tree, thirty to forty feet high, with straight stem,
rarely more than a foot in diameter. Wood white, very hard
and heavy, used for making beetles for splitting rails, mallets,
mauls, and similar implements. Bark on old trees dark-brown,
and furrowed, not smooth as in the closely allied Water Beech
(*Carpinus*). Seeds ripen in August in our Northern States, at
which time the hop-like catkins containing them should be
gathered and spread out to dry in the shade, until the seed can
be rubbed or threshed out. A handsome tree, well worthy of
extensive cultivation for its valuable timber. More or less
common in Nova Scotia, Canada, and all of our Northern States,
and in rich woods south to Florida. The European Hornbeam (*O. vulgaris*) resembles our native species very closely, and
is often planted for ornament.

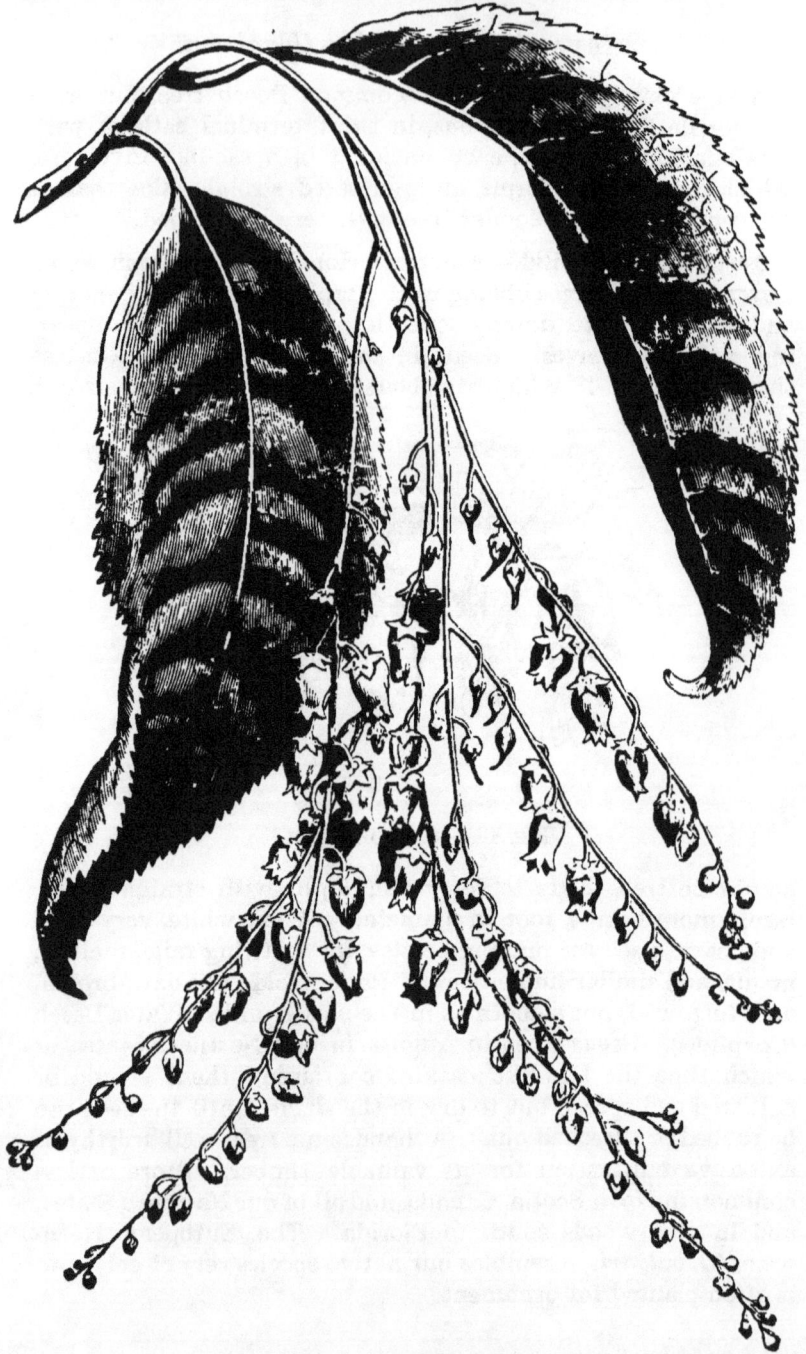

Fig. 44.—SORREL TREE.

OXYDENDRUM.—*Sorrel Tree.*

A beautiful little native tree belonging to the Heath Family, with leaves resembling those of the common peach tree, but a little larger. They have an acid taste, hence both the generic and common names refer to the sour taste of the leaves. There is only one species.

Oxydendrum arboreum, DC.—Sorrel Tree.—Leaves smooth, oblong-lanceolate, pointed serrate, on slender petioles. Flowers small, white, in one-sided racemes as shown in fig. 44. The racemes are clustered in loose panicles at the end of the branches, appearing late in spring. A rare tree in cultivation, probably because difficult to propagate, except from seed, and the seedlings make a slow growth for the first few years. The leaves change to a brilliant light crimson color in early autumn, and remain on the trees until cut by severe frosts. The wood is quite hard and fine-grained, but has not been sufficiently abundant to attract much attention. A small tree, but sometimes reaches a hight of fifty or sixty feet. Hardy in my grounds and probably farther north.

PARKINSONIA, Linn.

A small genus of about eight species, one-half of which are natives of North America, but in the warmer regions. Leaves large and much divided, the leaflets bipinnate. Fruit, long bean-like pods, containing several seeds.

Parkinsonia aculeata, Linn.—Leaves twelve to eighteen inches long, with small, but numerous spiny leaflets. Flowers yellow, in axillary racemes, three to six inches long. Pods two to ten inches long, containing from one to five seeds, the pods being contracted or compressed between the seeds. A small tree in Texas, and through Mexico. Wood hard.

P. floridum, Watson.—Somewhat larger leaflets than the last, with axillary racemes, the pods narrow with acute margins on the ventral side, seeds also thinner. A small tree on the Rio Grande, Southern Texas.

P. microphylla, Torr.—Leaflets few and pinnate, quite short, four to six leaflets in each. Flowers deep straw-color, the upper petals white. Pods two to three inches long, and one to three-seeded. Southern Arizona.

P. Torreyana, Watson.—Leaves composed of two to three pairs of leaflets, oblong, narrowed towards the base. Flowers in long racemes at the ends of the branches, bright yellow. Pods two to three inches long, containing two to eight seeds. The pods but slightly contracted between the seeds. A small tree twenty to thirty feet high, with light-green, smooth bark. This is the *Palo verde* of the Mexicans or Green-bark Acacia. Wood hard, and much valued for fuel.

PAULOWNIA, Siebold.

A noble Japanese tree introduced into this country nearly forty years ago, and has long been a popular ornamental tree on account of its large tropical-looking leaves, and handsome fragrant flowers. Grows freely from cuttings of the roots or seed. There is but one species.

Paulownia imperialis. — Imperial Paulownia. — Leaves large, heart-shaped, resembling those of the catalpa, but usually much larger, and on young, thrifty shoots, they are frequently one to two feet broad. Flowers trumpet-shaped, in large, upright branching panicles, violet color, and fragrant. A very rapid growing tree when young, but after reaching a hight of twenty or thirty feet, the branches spread laterally to a great distance, forming a rather broad, flat head. Hardy at the North, but in the latitude of Boston and Central New York, the flower buds are frequently killed in winter.

PERSEA, Gærtn.—*Red Bay.*

A small genus of evergreen trees and shrubs of the Laurel Family. Flowers greenish or white, and the fruit a small ovid drupe. Two species natives of our Southern States, and one in the West Indies.

Persea Carolinensis, Nees.—Red Bay.—Leaves oblong or lanceolate, smooth, two to three inches long, deep-green above, whitish beneath. Flowers silky, in roundish clusters, on short stems. Fruit deep blue. A large tree, forty to seventy feet high in rich, shady woods of North Carolina to Florida, and westward along the coast to Texas. Wood reddish or rose-color, hard, strong, durable, and susceptible of a high polish.

P. Catesbyana. — Michx.—Leaves smooth, lanceolate-oblong, sharp-pointed. Flowers minute, white, and somewhat downy within. Fruit small, black, on club-shaped stalks. A small tree, but more often a low shrub. Southern Florida.

PINCKNEYA, Michx.—*Georgia Bark.*

A genus of one, or at most, two species of small evergreen trees, closely related to the *Cinchona,* which yields the well known Peruvian bark. First made known by the elder Michaux, who found the trees growing on the St. Mary River in Florida in 1791, and carried seeds and plants to Charleston, S. C., and planted them in his garden near that city, where they had reached a hight of twenty-five feet in 1807, as stated by his son in his great work, North American Sylvia, vol. I, p. 180.

Pinckneya pubens, Michx.—Georgia Bark.—Leaves large, oval or oblong, smooth above, hoary pubescent underneath. Flowers tubular, an inch and a half long, white, with broad stripes of pink on the tube and in the center of the revolute petals. Fruit a globose papery, two-celled, capsule, opening at the top and containing numerous small seeds. A small tree with a wide spreading top, seldom more than twenty-five feet high, or stem over six inches in diameter. Wood very soft, and of no value, but the bark has been used more or less as a substitute for Peruvian bark, as it contained similar bitter tonic properties. Found wild on the marshy banks of streams in South Carolina and Florida.

PIRUS, Linn.—*Apple, Pear, Etc.*

An extensive genus, containing about forty species, principally in the temperate regions of Europe, Asia, and America. The apple, pear, crab apple, quince, service tree, mountain ash, and their many varieties, are all included in this genus. As there are few trees among them worthy of the arboriculturist's attention, I shall omit all except those inhabiting the United States.

Pirus Americana DC.—American Mountain Ash.—Leaves composed of thirteen to fifteen lanceolate, taper-pointed serrate leaflets. Flowers white, in large, flat cymes or 'clusters. Fruit in large clusters, not larger than peas, bright-scarlet, remaining on the tree until winter. A handsome ornamental tree, twenty to thirty feet high, reaching a very high northern latitude, even being found in Greenland and Labrador, and throughout the Canadas, all of our more Northern States, and southward along the mountains to North Carolina. There are several cultivated varieties of this species, also a very large number of the European Mountain Ash (*P. aucuparia*), which may be found described in nurserymen's catalogues.

P. angustifolia, Ait.—Narrow-leaved Crab Apple.—Leaves lanceolate or oblong, acute at the base, serrate. Flowers few in a cluster, rose-color, very fragrant. Fruit very acid. A small

tree, said to be found in Pennsylvania and southward, but I am
inclined to think not very common, as I have failed to find it
in cultivation, or obtain specimens from my correspondents
who reside in the regions where it is said to be indigenous.

P. arbutifolia, Linn.—Choke Berry.—Leaves oblong or obovate,
finely serrate. Flowers white or tinged with purple. Fruit
pear-shaped or round, red, sometimes purple. There are several
wild varieties, one with black fruit. This is the *Aronia arbutifolia* of Ell. A small tree or large shrub, sometimes ten or
twelve feet high. In swamps South.

P. coronaria, Linn.—American Crab Apple.—Leaves simple on
long, slender petioles, ovate or roundish, very smooth, and two
to three inches long. Flowers few in a cluster, rose-color, and
very fragrant. Fruit an inch or more in diameter, rather
broad and flat. Very acid and astringent; usually of yellowish-
green color. A small tree, but in rich alluvial soils sometimes
twenty-five feet high. Wood light-colored, but hard and fine-
grained. A handsome ornamental tree. Central New York,
west to Wisconsin, south along the mountains, and in the Mississippi Valley.

P. rivularis, Dougl.—Oregon Crab Apple.—Leaves simple ovate-
lanceolate, acute or acuminate, one to three inches long, sometimes three-lobed, more or less woolly-pubescent, as well as the
young branches. Flowers small, white. Fruit red or yellow,
about a half inch long. A small tree twenty to twenty-five
feet high. In low grounds in California and northward to
Alaska.

P. sambucifolia, Cham. & Schlect.—Western Mountain Ash.—
Leaves pinnate, and leaflets in four to six pairs, oblong-acute,
sharply serrate. Flowers white, like those of the Eastern
Mountain Ash. Fruit red, round, and about a quarter of an
inch in diameter. A small shrub. In the Sierra Nevada, and
north to Sitka.

PISCIDIA, Linn.—*Jamaica Dogwood.*

A small genus of tropical trees with unequal pinnate leaves,
and pea-shaped flowers in terminal or axillary spikes. Fruit a
bean-like pod, contracted between the seeds. We have one species.

Piscidia Erythrina, L.—Jamaica Dogwood.—Leaflets seven to
nine, oblong-ovate, abruptly pointed. Young branches, leaves
and flower-stalks silky and whitish, but becoming smooth with

age. Flower panicles, axillary or terminal, white with red veins. Leaves deciduous. Pods about two inches long. Nuttall says the wood is heavy, hard and resinous, light-brown, rather coarse-grained, but durable. A small tree in Southern Florida and through the West Indies.

PISTACIA, Will.—*Pistacia Nut, Etc.*

A small genus of diœcious sub-tropical trees, mostly natives of Southern Europe. One, the *P. officinalis*, is extensively cultivated in Sicily, for its fruit known as the Pistacia nut. Another, the *P. Terebenthus*, yields the Cypress turpentine, while the *P. lentiscus* produces a mastick, much used among the Armenian women for cleaning the teeth and perfuming the breath. One species in North America.

Pistacia Mexicana, HBK.—Mexican Pistacia.—Leaves composed of from five to ten pairs of small, oblong-ovate leaflets, on a slightly winged leaf-stalk. Flowers diœcious and without petals, in axillary clusters or panicles. Fruit small, somewhat compressed. A small tree in Western Texas, Southern California, and through Mexico.

PITHECOLOBIUM, Martin.—*Cat's Claw.*

The genus *Inga*, from which the *Pithecolobiums* have been separated, is an extensive one, and the species may be truthfully said to encircle the entire globe in tropical climates. They are mostly trees of large size. They are all evergreen, with acacia-like foliage. Fruit a legume or bean-like pod. We have one species, the

Pithecolobium Unguis-Cati, Benth. — Cat's Claw. — Branches usually spiny, but sometimes unarmed. Leaves bi-pinnate, leaflets four ; thin and obliquely obovate. Flowers yellow in globose-heads, in a loose raceme. Pods spirally twisted, containing five to six white seeds. A small tree fifteen to twenty feet high in Southern Florida and the West Indies.

PLANERA, Gmelin.—*Planer Tree.*

A small genus of deciduous trees, closely related to the elms, but with nut-like wingless fruit. All are handsome ornamental trees.

Planera aquatica, Gmel.—Planer Tree.—Leaves from an inch to an inch and a half long, ovate, in short petioles, sharp-pointed, serrate, with a rough surface. Flowers in small clusters, appear-

ing before the leaves. Seed ovate, covered with warty scales.
A small tree forty to fifty feet high, in low, moist soils, in
North Carolina and southward. Not quite hardy north of Phila-
delphia, except in very favorable sheltered situations. The
Japanese species or Kiaka Elm (*P. acuminata*) is a far more
hardy and robust growing tree than our native species. It has
large, glossy, smooth leaves, on red stems. The young shoots
are also red. A handsome and desirable ornamental tree. The
Caucasian Planera, *P. parvifolia*, also thrives very well in New
York and vicinity.

PLATANUS, Tour.—*Buttonwood, Sycamore, Etc.*

A genus of about a half dozen species, all but one inhabiting
North America. Large trees, with very close, smooth bark,
which, as the stem and branches enlarge, breaks up and falls
off in large flakes. The flowers are in dense, globose, naked,
unisexual heads, mingled with minute hairy scales, forming a
dry, rough, one-celled, and one-seeded fruit, pendulous, and
usually remaining on the trees until late in winter. All the
species and varieties may be readily propagated with ripe wood
cuttings of either one or two-year-old wood, but they should
always be planted in a moist soil.

Platanus occidentalis, Linn.—Buttonwood, Sycamore.—Leaves
large, six to ten inches broad, roundish heart-shaped, but deeply
and angularly lobed and toothed, covered when young with
dense whitish down, but soon becoming smooth. The pendulous fruit about an inch in diameter. One of the largest trees
found east of the Rocky Mountains, often from seventy-five to
a hundred feet high, with stem ten to fifteen feet in diameter.
The stems of large specimens often becoming hollow, only a
shell of three or four inches in thickness remaining sound.
These old hollow trunks were utilized by the early settlers in
Western New York, Ohio, and Indiana, for grain bins, smoke-
houses, and similar purposes, and then sometimes the pioneer
and his family found shelter in them, for it was an easy matter
to cut down one of the large, hollow trees, and then divide it
into sections of the required size with saw or axe. A few
slabs or pieces of bark, or slab-like sections of the same tree,
made a good roof or cover. Wood brownish, cross-grained,
cannot be split, and for this reason is in demand for meat-
blocks and similar purposes. Decays quickly if exposed to the
weather. Common in all of our Northern States, and southward

to Florida. It grows to a large size, and is very abundant in the bottom lands along the Ohio and Mississippi Rivers.

P. racemosa, Nutt.—California Buttonwood.—Leaves very variable, densely downy when young, broadly heart-shaped in outline, three to five-lobed, usually above the middle, lobes sharp-pointed, entire, or coarsely toothed. The leaves often a foot broad, and sometimes two feet on young, thrifty sprouts. Fertile heads two to seven in a string, a necklace-like spike. Fruit an inch in diameter. A common tree from the Valley of the Sacramento to Southern California. Bark very white, wood brittle, but said to receive a good polish and to be more durable than that of the Eastern species. The largest tree, whose measurement has been reported, is growing in Santiago Canyon, Los Angeles County, and was found to be twenty-nine feet and seven inches in circumference, but the trees rarely reach a hight of a hundred feet, or more than six feet in diameter (Botany of California, Vol. II.) *Platanus Wrightii*, Watson, is closely allied to the above, and probably Mexican, but said to be found in Southeastern Arizona.

FOREIGN SPECIES AND VARIETIES.

The Oriental Plane tree, or Sycamore, is better known in cultivation than our native species, because it was early introduced and more extensively propagated by our nurserymen. By some persons it is considered more desirable as an ornamental tree, as it has a more graceful habit. The branches are not quite as rigid, but often curved downward. In some instances it is decidedly drooping, with the ends curving upward. Some authors recognize two Asiatic species, the *P. orientalis*, Linn, and *P. cuneata*, Willdenow, but it is doubtful if they are distinct species. There are many varieties in cultivation, among which I will name *P. umbraculifera*, a dwarf, tortuous growing variety. *P. acerfolia*, the maple-leaved. *P. nepalensis*, with cut leaves and a pyramidal habit. *P. liriodendrifolia* has leaves resembling the Tulip Tree, raised by a nurseryman near Meton, Italy. *P. quinquelobata*, a variety with leaves divided into five-lobes. *P. asplenifola*, the leaves of which are very evenly and symmetrically divided. All the species and varieties of Platanus are hardy in our Northern States, at least so far as they have been tested.

POPULUS, Tour.—*Poplar, Aspen.*

A genus of about twenty species of deciduous trees, one-half the number natives of North America, and the others in-

habitants of the Old World, mainly in the northern or colder
regions. Flowers minute, in drooping catkins, on a cup-shaped
disk, usually appearing before the leaves. Seeds minute, usually furnished with a long tuft of cottony down at one end.
Trees of light, soft wood, of little value except for fuel, and
for this reason only prized where better kinds are scarce or un-
attainable. They are usually propagated from cuttings, as
most of the species can be rapidly multiplied in this way. The
rapidity with which some of the larger species grow, has made
them very popular for planting in the prairie regions of the
West, and while they have, no doubt, served a good purpose,
they are at the same time far inferior to many other kinds of
our indigenous forest trees. Some of the species have been described under quite a number of different names, all of which
will be found elsewhere, only one being employed in connection
with my remarks on each.

Populus angustifolia, James.—Willow-leaved Poplar.— Leaves
three to four inches long, taper-pointed, slightly heart-shaped
at base, serrate, smooth, shining, bright green. Branches rather
slender with smooth bark. Usually a stocky tree with a broad,
open, rather graceful head, forty to sixty feet high, with stem
two to three feet in diameter. Bark on old trees thick and
deeply furrowed. Wood light-colored, soft and spongy, of little
value. A handsome tree, resembling a Willow more than the
ordinary Poplars of the East. Common in the canyons of Arizona, Northern New Mexico, Colorado, and northward to the
Columbia River.

P. balsamifera, Linn.—Balsam Poplar, Tacamahac, Balm of
Gilead.—Leaves ovate, gradually tapering and pointed, sometimes heart-shaped, finely serrate, smooth on both sides.
Branches round, buds large and covered with a fragrant resinous matter, which appears to become volatile on the approach
of warm weather, and is widely diffused. There are several
natural local varieties of this species, among which are var. *P.
candicans*, Gray, *P. nigra*, Catesby, etc., etc. A tall, rather
pyramidal-shaped tree, along the banks of streams from Wisconsin to New England, and northward to the Arctic regions.
A rare tree in forests except far north, but has long been a
favorite ornamental tree for planting near dwellings, probably
on account of its odoriferous buds, which are supposed to
possess valuable medicinal properties, and are often gathered
and used for making an ointment that has a good reputation

among the people in the country in the treatment of wounds, bruises, rheumatism, and tumors. The name, "Tacamahac," was given this tree on account of the resemblance of the balsamic coating of the buds to the genuine *Tacamahaca*, or resinous product of *Fagara octandra*.

P. Fremontl, Watson.—Fremont's Poplar.—Leaves broadly triangular, or somewhat kidney-shaped, with a broad, acute point, with only a few serratures in each side. Leaf-stalks one to two-and-a-half inches long. Fruiting catkins three or four inches long; seeds small, white. A large tree, with gray, cracked bark, that on the young branches yellowish. Twigs round, smooth, not winged or angled. Along the Sacramento River in California, and westward to Utah. Var. *Wislizeni*, Watson, has sharply acuminate leaves, with very slender pistillate catkins, two to six inches long. This is the *P. monilifera*, Torr., in Botany of the Mexican Boundary Survey, and found further south than the species.

P. grandidentata, Michx.—Large-toothed Aspen.—Leaves three to five inches long, roundish-ovate, with large, irregular, sinuate teeth, and when young, densely covered with white, silky wool, but becoming smooth on both sides. A large tree sixty to eighty feet high, with rather smoothish gray bark. Wood light and soft, and of late years used for paper pulp. Common in the north, from Nova Scotia, Canadas, and the Northern States, but rare southward except along the Alleghanies.

P. heterophylla, L.—Downy-leaved Poplar. — Leaves heart-shaped or roundish-ovate, with obtuse, incurved teeth; white, woolly when young, but becoming smooth, except on the elevated veins beneath. Branches round. A large tree seventy or eighty feet high, not common or very abundant. New England to Illinois; southward to Arkansas, and eastward to North Carolina.

P. monilifera, Ait.—Cottonwood, Carolina Poplar.—Twigs and smaller branches thick, smooth, but sharply-angled or winged. Leaves large, six to nine inches long, broadly heart-shaped, smooth, sharply serrate, with slightly incurved teeth. Fertile catkins very long, with scales finely fringed, but not hairy; a very large tree, often a hundred feet high, with stem four or five feet in diameter. Wood soft, light, but burns rapidly when seasoned, but gives out little heat. A common tree in moist, low grounds, from New England to Colorado and Idaho,

and southward to Florida, but most abundant in the Mississippi
Valley. This species has been extensively planted on the prairies, and is still highly recommended as a forest tree, but its
merits consist mainly in the facility with which it is propagated, and rapidity of its growth, the wood being very inferior,
even for fuel, to some of the other species of this genus. There
is a handsome golden-leaved variety of this species, also a weeping variety, both handsome little trees.

P. tremuloides, Michx.—Quaking Asp, American Aspen.—
Leaves roundish, heart-shaped, with a sharp point, and somewhat regular teeth, smooth on both sides, with downy margins.
The leaf-stalks long and slender, slightly flattened on the sides,
which probably accounts for the constant trembling of the
leaves, when there is the slightest breeze. A common and
well-known tree, both in forests and under cultivation. A
widely distributed species, extending entirely across the Continent, through British America to the Pacific, extending northward to the Arctic Ocean. Usually in dense groups on moist
soils, on high elevations in our mountain ranges. I have found
large groves of this species in the Rocky Mountains at an
elevation of ten thousand feet. A medium-sized tree, fifty to
seventy-five feet, with stem twelve to twenty-four inches in
diameter. Bark smooth, hard, and thin, whitish on the outside. yellow within, quite brittle. Wood white, soft, but of
a firm texture, somewhat resembling that of the White Birch,
makes good fuel, and a Quacking Asp log will hold fire longer
than any other kind of wood I ever tried while camping in the
Rocky Mountain regions. The Indians are well acquainted
with this property of the Quacking Asp, and in moving their
camps, they use a brand or coals of this tree for taking fire
from the old to the new. It is also a favorite tree with the
beavers for building their dams.

P. trichocarpa, Torr. and Gray.—California Balsam Poplar.—
Leaves heart-shaped, or ovate to lanceolate, scalloped with
rounded teeth, two to four inches long; stalks an inch or two
long; fertile catkins five or six inches long. Seeds nearly
white. A large tree in California, from San Diego northward
to British Columbia. In Washington Territory, it is said to
grow nearly one hundred feet high, with stem three to six feet
in diameter. In low valleys and canyons near streams.

FOREIGN SPECIES AND VARIETIES.

I do not know of any species or varieties of the exotic Poplars, that are of any economic value, although there are a few worthy of cultivation as ornamental trees. The two foreign species best known in this country, are :

Populus alba.—Abele Tree, or Silver Poplar.—A large, rapid-growing tree, native of Europe, with large-lobed leaves, green above and silvery white beneath. A rather handsome tree, but a decided nuisance, owing to the great abundance of suckers, which come up almost constantly from the roots. There are several varieties of this species, but all have the same habit, and for this reason are not to be recommended, except for planting in locations where the suckers will not interfere with the growth of other kinds of plants. If grafted on stocks of species that do not throw up suckers, the Silver-leaved Poplars might be admitted into grounds of limited extent.

P. dilatata, or fastigiata.—Lombardy Poplar.—A century ago this was a favorite tree for planting near churches, cemeteries, and dwellings, but of late it is seldom employed, except to give variety of form in arranging the trees planted in large parks and pleasure grounds.

P. nigra.—European Black Poplar.—Has wide, spreading branches, with very large leaves, and very sticky or glutinous buds.

P. suaveolens.—This species is from Central Asia, and was introduced by Dr. Regel, director of the Imperial Gardens at St. Petersburg, Russia.—It possesses a very agreeable aromatic odor ; hence the specific name. This is very evident when the buds are rubbed between one's fingers.

There is a handsome weeping variety of this (*P. nigra pendula*), that is propagated by grafting high upon strong stocks of some erect young species. In addition I may name the Weeping Grecian Poplar (*P. græca pendula*). The Curled-leaved Poplar (*P. crispa*). Several additional species and varieties have been introduced from Europe and Asia, and may be found described in nurserymen's catalogues.

PROSOPIS, Linn.—*Mesquit, Screw Bean.*

A genus of nearly twenty species of tropical-evergreen, spiny trees, closely allied to the *acacias*, having pinnate leaves, pea-shaped flowers, and fruit a bean-like pod, containing several seeds. About a dozen species inhabit South America, and northward to Mexico, two only extending into the United States.

Prosopis juliflora, DC.—Honey Mesquit, Algaroba.—Leaves composed of from six to thirty pairs of short oblong, or linear leaflets, a half inch to an inch and a half long. Flowers very minute, greenish-yellow, in cylindrical spikes. Pods six inches or more in length, straight, or somewhat curved, contracted between the seeds. These pods, at certain stages of ripeness, are pulpy, and the pulp is quite sweet and sugary. Branches and twigs armed with short, strong spines. In figure 45, is shown a branch of Mesquit with leaves, flowers, and a pod, of nearly natural size. The Mexicans and Indians make use of the bean, out of which they form a kind of meal called *pinole*, and although of a sweet, nauseous taste to the civilized palate, it is considered wholesome. These beans are also fed to horses, cattle, and they are quite a luxury to the Donkey or Mexican Burro. The tree exudes a clear gum, very much like Gum Arabic, for which it may some day become a substitute, at least for many purposes. The Mesquit is a small tree, seldom growing more than thirty feet high, and more often it is a straggling shrub. Sap-wood yellowish, heart-wood reddish-brown, very hard and durable, making a most excellent fuel, and for this purpose superior to the best hickory, and has long been employed in smelting furnaces in Mexico and Arizona. It is found on the plains of Western Texas, New Mexico, Southern Colorado, Southern California, and southward through Mexico.

P. pubescens, Benth.—Tornilla, Screw-pod Mesquit.—Leaflets in five to eight pairs, oblong, very short. Flower spikes one to two inches long. Pod thick, spirally twisted, with numerous turns, forming a narrow, straight cylinder one to two inches long, pulpy within. The pods of this species are also ground into meal by the Indians, and fed green or when nearly ripe to their ponies. A shrub or small tree, twenty to thirty feet high. Wood similar to the last. In San Diego, California, near Fort Mohave, and east to New Mexico, and southward through Mexico.

PRUNUS, Tour.—*Plum and Cherry.*

An extensive genus of about eighty species, distributed over the northern hemisphere, mainly in temperate climates. The genus includes many of our best known, and best cultivated

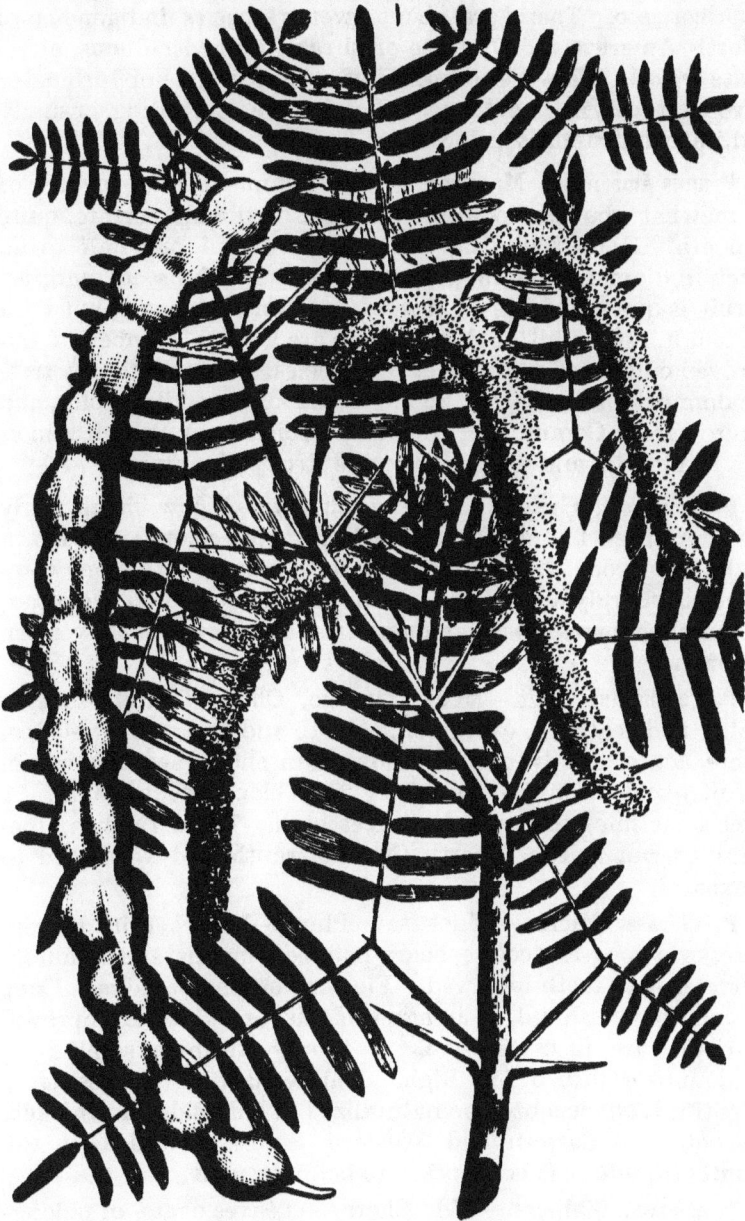

Fig. 45.—MESQUIT TREE.

fruits, the Plum, Cherry, Sloe, Cherry-laurel, Peach, Nectarine, Apricot, etc. There are about twenty species indigenous to North America, mostly trees or shrubs, with deciduous, alternate leaves. Flowers composed of a campanulate or turbinate, five-cleft calyx, deciduous. Petals five, spreading, usually white or but slightly colored,

Prunus Americana, Marshall.—Wild Plum.—Leaves ovate, or somewhat obovate, pointed, coarsely or doubly serrate, quite smooth. Fruit roundish, oval, yellow or red, one half to an inch in diameter, having a flattened stone with broad margins. Fruit is quite variable in flavor, sometimes pleasant, but with a tough, rather bitter skin. There are a large number of improved cultivated varieties of this species. A small, thorny tree, seldom over twenty feet high. Wood of a reddish color, and quite hard. Common in low, moist soils, from British America to Florida, and westward to the Rocky Mountains.

P. Andersoni, Gray.—Anderson Cherry.—A low shrub, only two or three feet high, with solitary, rose-colored flowers, a half inch broad, fruit small, and thin fleshed, with stone compressed, sharply angled on one side, and furrowed on the other, resembling a small peach stone. In foot hills of Northwestern Nevada.

P. Caroliniana, Ait.—Mock Orange, Cherry-Laurel.—Leaves thick and leathery, evergreen, smooth and glossy, ovate-lanceolate, acute, mostly entire. Flowers in short racemes, white. Fruit ovoid, soon becoming dry and black, stone round. A tree sometimes thirty or forty feet high. Wood reddish, fine-grained, but brittle. North Carolina south, and westward to Texas.

P. Chicasa, Michx.—Chickasaw Plum.—Leaves thin, lanceolate, or oblong-lanceolate, sharp-pointed, smooth, and minutely serrate, with teeth incurved. Flowers on short stalks. Fruit round, yellowish-red, of an agreeable flavor. Several improved varieties are in cultivation. A thorny shrub or small tree, seldom over fifteen feet high. Native habitat not positively identified, but has become naturalized in old fields and thickets in both the Eastern and Western States, as well as in the Southern, where it is supposed to be indigenous.

P. demissa, Walpers.—Wild Cherry.—Leaves ovate, or oblong-ovate, usually broadest above the middle, abruptly pointed, rounded or heart-shaped at base. Flowers white, in terminal racemes, appearing after the leaves. Fruit small, round, purplish-black, or red, sweet and edible, but somewhat astringent. An erect, slender shrub, three to twelve feet high. San Diego northward to the Columbia River, and eastward to the Rocky Mountains. I found this species with several natural varieties growing in the canyons of New Mexico, at an elevation of seven or eight thousand feet. The fruit on some plants are quite small and black, on others near by large or nearly a half inch in diameter, and of a bright red color, and produced in long, drooping racemes. I have some very promising seedlings of the large red variety, which I hope to fruit soon.

P. emarginata, Walpers.—California Cherry.—Leaves oblong-obovate to oblanceolate, one to three inches long, narrowed to a short petiole, with one or more glands near the base. Flowers six to twelve in a cluster. Fruit round, black, about one-third of an inch long, very bitter and astringent. A small shrub four to eight feet high.

Var. *mollis*, Brewer, is said to be a much taller-growing, reaching a hight of twenty-five feet, mostly in open forests in Northern California, Oregon and Washington Territory.

P. fasciculata, Gray.—Dwarf Cherry.—Leaves small, a half-inch long, in bundles or clustered, obtuse or acutish, with very short stalks. Fruit very small, hairy or velvety skin ; pulp thin, stone acute at both ends, smooth and scarcely margined. A small, much-branched shrub, two or three feet high. In Southern Sierra Nevada, Utah and Arizona.

P. ilicifolia, Walp. — Evergreen or Holly-leaved Cherry. — Leaves thick and rigid, shining above, broadly-ovate, obtuse or acute, somewhat heart-shaped at base, spinosely toothed, an inch or two long. Flowers small in racemes, as shown in fig. 46. Fruit large, half an inch in diameter or more, usually red, but often dark purple or black. Pulp acid and astringent, but pleasant flavored. The bark is gray, rather rough. Wood close-grained, tough and of a reddish color. A large and handsome evergreen shrub in the Coast Ranges of California, from San Francisco to San Diego and Western Arizona.

P. maritima, Wang.—Beach Plum.—Leaves ovate or oval, finely serrate, soft, velvety underneath. Flowers white, produced in great abundance. Fruit globular, dark purple, about a half inch in diameter, edible and rather pleasant flavored. Under cultivation, the fruit becomes much larger than in the

9

wild state. A low, straggling, thorny shrub, six to ten feet high, along the sea-beach from Massachusetts, southward to Virginia. A handsome ornamental shrub that thrives inland, and in dry, sandy soils, as well as on the sea-coast.

P. Pennsylvanica, L.—Wild Red Cherry.—Leaves oblong-lanceo-

Fig. 46.—HOLLY-LEAVED CHERRY.

late, pointed, finely serrate, shiny, green and smooth on both sides. Flowers many in a cluster, on long stems. Fruit round, light red, quite small, pulp thin and quite acid. A small tree,

twenty to thirty feet high in rocky woods, often taking possession of abandoned woodlands, or those from which the trees have
been destroyed by fires or tornadoes. This species of wild
cherry has been recommended as a stock for the cultivated varieties ; but I am not aware that it has been used to any considerable extent by nurserymen. A common tree far to the
north, and along the mountains southward to North Carolina.

P. pumila, L.—Dwarf or Sand Cherry.—Leaves obovate-lanceo-
late, tapering to the base, somewhat toothed near the apex.
Flowers small, white, few in a cluster. Fruit one-fourth to a
half-inch in diameter, dark red. Flesh of a sub-acid or rather
insipid taste. A low spreading or prostrate shrub, with many
slender stems. A rather unproductive shrub in cultivation, but
wonderfully prolific when growing in the sands along the
shores of our northern lakes and ponds. Plants of this species
have often been offered for sale by tree-peddlers, in fact, large
numbers have been sold under such names as Utah Cherry,
Dwarf Cherry, etc., but it is not worth cultivating except as a
curiosity. It is found wild in Massachusetts, and westward to
Lake Superior.

P. serotina, Ehrh.—Wild Black Cherry.—Leaves oblong, taper-
pointed, serrate, with incurved short teeth ; rather thick,
smooth, and shining above. Flowers in long pendulous racemes.
Fruit purplish-black, slightly bitter, but with a pleasant vinous
taste. Wood light red, close-grained, easily worked, and long
known as one of our most valuable native woods for various
kinds of cabinet work. A large tree, sixty feet high and over,
with stem three to four feet in diameter. Once very abundant
in our Northern States, but trees of large size are becoming
quite scarce. It is found in forests as far north as Hudson's Bay,
and south to Florida, and west to Texas, and northward in the
Valley of the Mississippi to Iowa.

P. subcordata, Benth.—California Plum.—Leaves ovate, heart-
shaped, or wedge-shaped at base, obtuse or acute, sharply and
finely serrate. Young branches and leaves pubescent in spring,
becoming smooth in summer. Fruit about three-fourths of an
inch long, red, and edible ; stone acutely edged on one side. A
scraggy shrub, four to ten feet high. California and Oregon in
dry, rocky hills.

P. umbellata.—Ell.—Leaves thin, ovate-lanceolate, acute at
both ends, sometimes the upper ones are rounded at the base ;

finely and sharply serrate, smooth or soft-downy beneath.
Fruit round, nearly a half-inch in diameter, dark-purplish or
black, acid and slightly bitter. This species may be only a
southern variety of our northern Beech Plum (*P. maritima*).
A shrub or small tree in light, sandy soils. South Carolina and
Florida.

Of the foreign species and varieties, there are such a vast
number, that I cannot afford the space that would be required
to mention them all, however briefly, besides they are mostly
fruit or small ornamental trees of no especial interest to the
practical forester. The common Sweet Cherry of our gardens
is descended from the *Prunus Cerasus* of Europe, or may be
Asia, as its native country is not positively known, for it has
run wild all over Europe, as well as in our Eastern States. The
sweet varieties are separated in a class by themselves, under the
general name of *Bigarreau* cherries, while the more dwarf and
acid varieties are called *Morellos*. There are many handsome
ornamental varieties of each, both weeping, double-flowering.
China, Japan, Nepal, the Himalayas, and various countries in
Southern as well as Northern Europe, have given us numerous species and varieties of the genus.

QUERCUS, Linn—*Oak*.

An extensive genus of nearly two hundred and fifty species,
distributed throughout the temperate regions of Asia, Europe,
and North America. It includes both evergreen and deciduous
trees and shrubs, with alternate, simple, or pinnately-veined
leaves. Staminate flowers in slender, drooping catkins. Pistillate flowers, solitary, in clusters, or sometimes in spikes, sessile in a cup-like, scaly involucre, which enlarges into a rough
cup around the base of a single, one-seeded nut or acorn. The
cotyledons thick and fleshy, remaining underground in germination, like those of the common garden pea, not lifted above
the surface as in the bean. For more than a hundred years the
botanists of the world have been at work at this most difficult
genus, and while in a measure they have brought "order out
of chaos," and especially in our North American species, there
is still much to be done before the oaks of the world are scien-
tifically described and correctly classified. The great work of
F. Andrew Michaux & Son, on the American Oaks and other
trees, published in Paris 1810–13, under the name of "North
American Sylva," and later published in this country, will long

remain a monument to the industry and scientific attainments
of the authors, but recent discoveries, especially in the Rocky
Mountain regions and westward, has not only added many new
species of the oak, but has also made it necessary to revise some
of the earlier classifications of the members of this genus. The
late Dr. George Engelmann, of St. Louis, Mo., a most capable
botanist, devoted much time to the study of the oaks, and published an excellent paper on the subject in the Transactions of
the Academy of Science, of St. Louis, Vol. III, 1876, and also
elaborated the oaks of California for the Botany of California,
edited by Sereno Watson, issued as supplementary volumes of
the Geological Survey. I accept Dr. Engelmann's arrangement
of the species, but may add that he was well aware of the difficulties to be met in attempting this work, for in the paper referred to, in speaking of the many varieties of the Rocky
Mountain scrub-oak, he says : "If one oak behaves thus, why
not others ? Thrown into a sea of doubt, what can guide us to
a correct knowledge." Having spent many months among
these scrub-oaks, I am fully aware of the difficulties to be met
in trying to determine where a variety ends, and a species begins, consequently am more than willing to throw the responsi-
bility of separating them upon some one else.

Quercus agrifolia, Nee.—Encino Holly-leaved Oak.—Leaves
oval to oblong, two to three inches long, usually obtuse or
heart-shaped at base, the uneven margins with spine-tipped
teeth, but these are sometimes absent. Petiole or leaf-stalks
downy. Acorns sub-sessile or sessile, solitary or in clusters,
maturing the first season, slender, and one to one and a half
inches long, and about one-third of an inch broad. This is one
of the Black Oaks. A large tree, with very thick gray bark,
and wood rather cross-grained and perishable. A very picturesque oak, with very stocky stem, sometimes twenty feet in
circumference, and Prof. Brewer reports specimens near Mount
Diablo, with a spread of branches of one hundred and twenty
feet. A variety (*Q. agrifolia*, var. *frutescens*), is only a shrub,
three to five feet high. A common tree in the maritime portion
of California.

Q. alba, L.—White Oak.—Leaves whitish, pubescent while
young, but soon become smooth, bright green above, with
three to nine oblong or linear-obtuse, mostly entire oblique
lobes. Leaves very persistent, many remaining on the trees all
winter, and only fall when pushed off by the expanding buds

in spring. Acorn ovoid-oblong, about an inch long, set in a
shallow, rough cup. The kernel sweet-tasted, or only slightly
bitter, edible. A large tree, sixty to eighty feet high, with stem
six feet and sometimes more in diameter. Wood light-colored,
heavy, very tough, and elastic, well-known as one of the most
valuable of American forest trees. The wood is very durable,
and is always in great demand for a variety of purposes, especially for agricultural implements, carriages, and ties for railroads. A common tree in our northern forests, extending southward to Florida. The white oak should be given a prominent
place in every collection of native forest trees cultivated for
economic purposes.

Q. aquatica, Catesby.—Water Oak.—Leaves perennial or evergreen, obovate-oblong, or wedged-shape, smooth on both sides ;
obtusely three-lobed at the summit, often entire, or on young
shoots, toothed or lobed, with bristle-like awns. Acorns small,
globular, downy, and set in a shallow saucer-shaped cup. A
small tree with smooth bark, seldom growing more than forty
feet high. Wood variable, sometimes tough, but more commonly rather brittle, used principally for fuel. In swamps and
along the banks of streams, from Maryland to Florida and westward.

Q. bicolor, Willd.—Swamp White Oak.—Leaves unequally and
deeply sinuate, toothed, almost pinnatifid, whitish, downy beneath, and bright green above. The leaves intermediate in
form, between the white and chestnut oaks, but the species is
usually classed with the latter. Acorns nearly an inch long,
oblong-ovoid, set in a shallow cup, often mossy-fringed at the
margin. A large tree, sixty to eighty feet high, and stem five
to eight feet in diameter. Wood closely resembling the white
oak, and valuable. Most common in the Northern and Western
States in moist soils, but also found South among the
mountains, but on moist or wet ground. Var. *Michauxii*, Nutt.,
has smaller leaves, and longer and more slender acorns. A
large tree in Southern Illinois, Delaware, Florida, and South
Carolina.

Q. Breweri, Engelm.—Brewer's Oak.—Leaves small, one and a
half to two or three inches long, deeply pinnatifid, lobes obtuse
and emarginate, sometimes again lobed on petioles. Acorns
sessile, an inch long, set in a shallow cup. A small shrub, two
to six feet high, on the middle or higher elevation of the Sierra
Nevada, from Calveras County, Cal., to the Oregon line.

Q. Catesbæi, Michx.—Turkey Oak, Scrub-oak.—Leaves rather thick and broad, narrowed into a short stalk, deeply lobed, the lobes very acute, from a broad base, six to nine inches long. Acorns rather large, but quite short, set in a thick turbinate cup an inch broad, the upper scales curved inward. A small, scraggy rough-barked tree in dry pine-barrens, from North Carolina to Florida.

Q. cinerea, Michx.—Upland Willow Oak.—Leaves entire, about three inches long, and less than an inch wide, obtuse or acute, white tomentose beneath, persistent, and almost evergreen. Acorns small, almost round, the cup enclosing about one-third of the nut. A small tree, twenty to thirty feet high in the pine-barrens of North Carolina and Florida. Michaux states that this species, like that of the Black Oak, affords a beautiful yellow dye, but the tree is too small to be of much value, even for fuel. Var. *pumila*, Michx. *Q. pumila*, Walt., is a low shrub, only two or three feet high, with lanceolate, wavy leaves. Fruit of species and variety biennial.

Q. chrysolepis, Liebm.—California Live Oak.—Leaves evergreen, oblong, acute, or terminating in a sharp, rigid point, obtuse or slightly heart-shaped at base, mostly entire on large trees, but on younger ones sharply toothed, sometimes both forms on the same branch, rather thick and about two inches long and an inch wide, yellowish, downy beneath, but after a year becoming bluish-white. Acorns oval, sometimes an inch and a half long, and only a half inch in diameter, set in a saucer-shaped cup, covered with triangular scales. Acorns maturing at the close of the second season. One of the largest oaks, with a flaky ash-gray bark in the Coast Ranges, and along the slopes of the Sierra Nevada. On the higher mountains it is often a mere shrub. Var. *vacciniifolia*, Engelmann, is a small shrub, three to six feet high, with acorns less than an inch long, with smaller leaves.

Q. coccinea, Wang.—Scarlet Oak.—Leaves long, petioled, oval or oblong, with deep and broad-scalloped edges, and six to eight entire or sparingly-toothed lobes, rounded at the base, smooth and shining on both sides. Cup top-shaped, enclosing about one-half of the roundish depressed acorn, which is usually from a half to three-fourths of an inch long. The leaves turn bright red or scarlet in late autumn, and are quite persistent, although dropping after severe freezing weather sets in. A handsome large tree with gray bark, rough, but not deeply furrowed,

Wood white, heavy, moderately coarse-grained, sometimes quite tough, but variable in texture and value. New England, and near the coast, southward, on sandy soils, on red-sandstone ridges of New Jersey and south. Engelmann says it is found in Minnesota. There is yet some doubt as to whether this is really distinct from *Q. tinctoria.*

Q. densiflora, Hook. & Arn.—California Chestnut Oak.—Leaves persistent, oblong-acute, obtuse or rarely acute at base, entire, with revolute margins, but sometimes slightly toothed, tomentose beneath, or whitish, two to five inches long, a half inch to two inches broad. Staminate flowers in erect catkins, with pistillate at base. Acorns biennial, oval or oblong, sharp-pointed, an inch to an inch and a half long, with a very thick shell set in a very shallow cup. Kernel very bitter. A species intermediate between the oaks and chestnuts. A middle-sized tree or shrub, but Professor Brewer says that in the Santa Cruz Mountains it grows to a hight of fifty or sixty feet, and rarely to eighty feet, with a stem two feet in diameter. Prof. Palmer reports it on the Coast Ranges of California, from the Santa Lucia Mountains, and among the red woods to the Shasta region.

Q. Douglasii, Hook. & Arn.—Mountain White Oak, or Blue Oak.—Leaves small, only an inch or two long, oblong-sinuate or with shallow lobes, sometimes almost entire, on short stalks, bluish-green, becoming smooth above, pubescent beneath. Acorns an inch or more in length, oblong, tapering or pointed, set in a shallow cup, covered with flat scales. A medium to large tree, with downy branchlets, on dry foot hills of the Coast Ranges of California, near the centre of the State. Resembles the white oak of the Eastern States, but does not grow as large. The largest trees seen by Professor Brewer had a circumference of nine feet.

Q. dumosa, Nutt.—Small-leaved Oak.—Leaves small, a half inch to an inch long, oblong-obtuse, rounded, or rarely acute at the base, entire or slightly sinuate on young shoots, toothed, dark green above and pubescent beneath. Acorns sessile, variable in size, an inch long or more, sometimes slender and small, set in deep cups, usually strongly tubercled. Var. *bullata* has leaves rounder, thicker and paler in color. A tall shrub or small tree, seldom over twenty feet high, and with slender straight branches. Leaves persistent through winter. In the Coast Ranges, from San Diego to San Francisco Bay. The variety in the Santa Lucia Mountains (Prof. Brewer).

Q. Emoryi, Torr.—Dwarf Evergreen Oak.—Leaves small, evergreen, slightly lobed, acorns very small. A widely spread shrubby evergreen oak in Southern Texas, New Mexico, Arizona, and southward in Mexico.

Q. falcata, Michx.—Spanish Oak.—Leaves oblong, rounded at the base, three to five-lobed ; the lobes entire or sparingly toothed at the apex, the terminal one commonly narrow and elongated. Acorns about a half inch long, set in a cup enclosing half of the roundish nut. A large tree, often sixty to seventy feet high, and stem four feet in diameter. Wood dark-brown or reddish, coarse-grained, decays rapidly when exposed to moisture. Bark thick, rich in tannin, and often extensively employed by tanners in making what is called "Oak-tanned Leather." New Jersey, southward to Florida, and westward to the Valley of the Mississippi. Very abundant in the Southern States.

Q. Carryana, Dougl.—Western Oak.—Leaves four to six inches long, by two to five wide on stalks, a half to one inch long, coarsely deeply cut-lobed ; lobes broad, obtuse, or sometimes sharp-pointed, dull green above, beneath pale-yellowish, and somewhat downy. Acorns sessile or on short stalks, one to one and a half inches long, oval, in small and very shallow cup. A large tree, seventy to a hundred feet high, and stem three to four feet in diameter. Wood said to be coarse, hard, and brittle. A common tree in the valleys north of San Francisco Bay, extending into Oregon and British Columbia.

Q. Georgiana, M. A. Curtis.—Georgia Oak.—Leaves three to four inches long, very smooth, somewhat obovate, and wedge-shaped at base, with deep or shallow sinuses, three to five-triangular or obtuse lobes. Acorn a half inch long, oval or roundish, set in smooth cups, enclosing one third of the nut. A small shrub, six to eight feet high, on Stone Mountain, Georgia.

Q. heterophylla, Michx.—Bartram Oak.—Dr. Engelmann places this among the hybrid oaks, and intermediate between *Q. Phellos* and *coccinea*, but Decandolle considered it a variety of *Q. aquatica*, which in some respects it certainly very much resembles, especially in the sharp-pointed lobes of its leaves. The original tree in the old Bartram Garden, Philadelphia, was long since destroyed, and was only a small tree, some thirty feet high at the time. But there are seedlings of it now twice that hight, differing somewhat from the original. At best, we may

say that this species is a doubtful one, although by some authors
it is thought that it is to be found in New Jersey, and southward to North Carolina.

Q. hypolenca. — Engelm.—Leaves thick, lanceolate, acute,
three or four inches long, and from three-quarters of an inch to
nearly an inch broad. Usually entire or slightly revolute on
the margin, but occasionally show one to three minute teeth.
Nearly smooth, and pale-green above, densely covered with yellowish down beneath. Acorns small, sessile, solitary or in pairs,
ovate, and set in a roundish cup, pubescent obtuse scales. A
small tree, sometimes thirty feet high, near the copper mines,
New Mexico (Thurber). Southern Arizona, at an altitude of
seven thousand feet (Rothrock), and also in Sonora, Mexico.

Q. imbricaria, Michx. — Shingle Oak, Laurel Oak.—Leaves
three to five inches long, lanceolate-oblong, acute or obtuse at
each end, tipped with an abrupt short point, pale-downy beneath, deciduous. Acorn globular, five-eighths of an inch long,
cup enclosing about one-third of the nut, scales broad, whitish,
closely appressed. A large, stocky tree, forty to fifty feet high,
with quite smooth bark even on old trees. Wood hard, heavy,
coarse-grained, easily split, and occasionally used for making
shingles of an inferior quality. A handsome tree when young,
the leaves resembling those of the chestnut. In barren and
open woodland, from New Jersey westward.

Q. illcifolia, Wang.—Bear, or Black Scrub-oak.—A low, dwarf
shrub, with leaves three to four inches long, obovate, wedge-
shaped at base, angularly, about five-lobed, white downy beneath. In sandy barrens and rocky hills. Acorns barely a half
inch long. New England to Ohio, and southward.

Q. Kelloggii, Newbury.—Kellogg's Oak.—Leaves deciduous,
thick, broadly oval, pinnatifid-lobed, the lobes tapering and
entire, or broad and lobed-dentate, at first downy, but soon becoming smooth, three to six inches long. Acorns oblong, over
an inch long, mostly on short stalks, one half to an inch long,
and several together ; cup round, but sometimes very deep,
with ovate-lanceolate imbricate scales. A medium or large-
sized tree with rough, black bark, in the Coast Ranges of California. For various synonymes see Index.

Q. laurifolia, Michx.—Laurel-leaved Oak.—Leaves three to
four inches long, oblong-lanceolate, entire or lobed, widest in
the upper third, or at least above the middle. Leaves persistent,

and remain on the trees until spring. Acorns biennial and quite small. A large tree, North Carolina to Florida.

Q. lobata, Nee.—Lobed-leaf Oak.—Leaves deciduous, two to four inches long, downy beneath, oblong or ovate, deeply-lobed, lobes sometimes toothed or lobe-dentate. Acorns elongated-conical, one to two and a fourth inches long, usually pointed. Cup deeply hemispherical, almost always strongly tubercula-ted. A large tree, with smooth, slender, and often pendant branches. Common throughout the State of California. Wood said to be brittle, and bark on old trees four or five inches thick.

Q. lyrata, Walt.—Over-cup Oak, Post Oak.—Leaves five to eight inches long, crowded at the ends of the branches, downy or pale beneath, narrowed at base, obovate-oblong, seven to nine lobes, the lobes triangular, acute and entire. Acorn round-ish, and nearly enclosed in the round-ovate cup with rugged scales. Acorns ripen the first season. A large tree in the swamps of North Carolina to Florida, and sparingly in Arkan-sas. Not very abundant. Wood said to be excellent, resemb-ling that of the White Oak.

Q. macrocarpa, Michx.—Burr Oak, Mossy-cup Oak.—Leaves large, eight to fifteen inches long, thin, obovate-oblong, slightly downy beneath, narrowed at the base, stalk short, slightly or strongly, and many-lobed, the lobes rounded and mostly entire. Fruit large, scales of the cup thick, the upper ones producing long, fringe-like awns. Acorn an inch to an inch and a half long, half enclosed in the cup A large tree, sixty to eighty feet high, with stem four feet or over in diameter. One is mentioned in Vol. I, North American Sylva, as growing in Ohio, with a stem seventeen feet in diameter, at six feet above the ground, and the tree one hundred feet high. The young twigs and branches are somewhat corky. Wood coarse-grained, of little value, except for fuel. A widely distributed tree in our Northern States, but not very abundant, except in the Western or from Ohio south and west.

Q. Muhlenbergii, Engelm.—Yellow Chestnut Oak.—Leaves thin, five to six inches long, one and a half to two broad, pale be-neath, sharply serrate, with incurved teeth, and either lanceo-late, with a long point, or broadly ovate or obovate, sometimes seven inches long and five wide. A small or medium-sized tree, with flaky, pale ash-colored thin bark, and very tough wood; light yellowish or brown when mature, whence probably the popular name of yellow oak. Occurs scatteringly throughout the Middle and Northern Atlantic States in Pennsylvania, only on limestone hills (Porter), but most abundant in tne Mississippi Valley. Wood valuable and very durable.

Q. nigra, L.—Black Jack Barren Oak.—Leaves large, five to ten inches, thick, broadly wedge-shaped, rounded at the base, mostly three-lobed at summit, bristle-awned, smooth above, and rusty-downy beneath, deciduous. Acorns biennial, one-half to three-fourths of an inch long, cup top-shaped, with coarse scales enclosing one-third or one-half of the oblong-ovate nut. A small tree, seldom more than twenty feet high, with very dark-colored rough bark. Wood coarse-grained, only valuble for fuel. · Widely distributed from New Jersey southward to Florida, and westward to Texas and northward.

Q. oblongifolia, Torr.—Evergreen White Oak.—Leaves evergreen, oblong, one to two inches long, and half as wide, on very short stalks, entire or with a few blunt teeth, obtuse at each end, or slightly heart-shaped at base, downy when young, calyx lobes short, oval, woolly. Acorns oblong, one-half to an inch long, cups hemispherical, tubercled. A small, handsome, evergreen tree, twency to thirty feet high, with stem two feet in diameter. Fruit maturing the first season. Mountains of Southern California and Mexico.

Q. Palmeri, Engelm.—Palmer's Oak.—A tall shrub, with thick and very rigid leaves, scarcely an inch long, round, oval, obtuse or sub-cordate at base, with undulate and spiny margins. Acorns maturing the second season. Mountains of San Diego County, California, near the boundary and southward.

Q. palustris, Du Roi.—Pin Oak, Swamp Oak.—Leaves oblong, smooth and shining, bright green on both sides, deeply pinnatifid, with broad and rounded sinuses, the lobes divergent, cut-lobed and toothed. Acorns globular, scarcely one-half inch long, cup shallow and saucer-shaped. A very handsome, medium sized tree, with light, elegant foliage, growing in low grounds, along streams, from New England to Nebraska and Kansas. Wood rather coarse-grained, but valuable for plank or for purposes where it will not be exposed to the weather.

Q. Phellos, L.—Willow Oak.—Leaves deciduous, linear-lanceolate, narrowed at both ends, two to three inches long, bristle awned, scurfy when young. Cup saucer-shaped, enclosing the base of the roundish nut. Acorn maturing the second year. A tree thirty to fifty feet high, with reddish, coarse-grained wood of little value, except for posts and beams in buildings, where it will not be exposed to moisture. A handsome ornamental tree. There are several varieties.

Q. Prinus, L.—Swamp Chestnut Oak.—Leaves ovate, oblong, or oblong-obovate, coarsely and somewhat dentate, with rounded teeth, downy beneath, and smooth above. Cup globular, or with a top-shaped base, thick, tubercled when old, nearly one half the length of the ovoid acorn, which is about one inch long, with a sweetish edible kernel; the acorns ripening the first year. A medium to large tree, with reddish, coarse-grained wood, much inferior to white oak. Vermont to Florida, and west to Mississippi, also west of the Alleghanies in Tennessee and Kentucky.

Q. prinoides, Willd.—Chinquapin Oak.—Mainly distinguished from Q. *Mühlenbergii* by its low stature and more undulate than sharp-toothed leaves, on shorter petioles, and commonly deeper cups. Dr. Engelmann says well enough marked eastward, but from Western Missouri to Kansas, it runs into the arborescent *Mühlenbergii*. A low shrub East, and a doubtful species.

Q. rubra, L.—Red Oak.—Leaves oblong, smooth, pale beneath, with eight to twelve entire or sharply toothed lobes. Leaves turning dark red after frost. Acorn an inch long, set in a shallow cup with fine scales. A very large and common tree, with reddish, very coarse-grained wood, but in some soils, moderately compact, and much used for hewn timber and staves for barrels, and similar vessels. Everywhere from Nova Scotia to Florida, and westward to Minnesota and Texas.

Q. stellata, Wang.—Post Oak.—Leaves four to six inches long, cut into five to seven roundish divergent lobes, the upper ones the largest and often notched; grayish downy underneath, and pale and rough above. Acorn about a half inch long, oval, cup encircling one third to one half the nut. A medium sized tree, forty to fifty feet high, with very hard, durable wood, resembling that of the white oak. Massachusetts to Florida, and westward to the prairies beyond the Mississippi Valley.

Q. tinctoria, Bartram.—Yellow-Barked Oak, Quercitron or Black Oak.—Leaves obovate-oblong, slightly or deeply lobed, the lobes sharply toothed, obtuse at the base, more or less rusty, pubescent when young. Acorns nearly round, one half to two thirds of an inch long, set in a rather deep, conspicuously scaly top-shaped cup. A large tree, sixty to eighty feet or more in hight, with reddish, close-grained, strong and durable wood, extensively employed by coopers and carriage makers. The bark is used in tanning, and long known as the quercitron of dyers. Very abundant in all the Atlantic States, but less common in the Western.

Q. tomentella. — Engelm.—Leaves oblong-lanceolate, two to three inches long, on short stalks, obtuse at base, acute or toothed, rarely entire, strongly ribbed with revolute margins, densely downy when young, becoming smooth. Leaves evergreen, or at least very persistent. Acorns ovate, and over an inch long, maturing the second season. A tree sometimes forty feet high, and closely allied to the California Live Oak (*Q. chrysolepis*), Guadalupe Island (Dr. Palmer).

Q. undulata, Torr.—Rocky Mountain Scrub-Oak.—One of the Rocky Mountain oaks that runs into almost innumerable forms, and from low, almost trailing shrubs, up to trees twenty or more feet high. The species has a general resemblance to the White Oak of the Eastern States, and the leaves are fully as persistent, and the wood is very hard and tough, but usually too small for any use except stakes and firewood. The leaves resemble those of the White Oak (*Q. alba*), only they are very much smaller. Dr. Engelmann refers to a few of the best known forms of this species, as follows : 1. Var. *Gambelii* (*Q. Gambelii*, Nutt., and probably *Q. Drummondii*, Liebm.) 2. Var. *Gunnisoni* (*Q. alba*, var. *Gunnisoni*, Torrey). 3. Var. *Jamesii*, Engelm. 4. Var. *Wrightii*, Engelm., often confounded with *Q. Emoryi*. 5. Var. *pungens*, Engelm., (*Q. pungens*, Liebm.) Var. *oblongata*, Engelm., (*Q. oblongifolia*, Torr.) 6. Var. *grisea*, Engelm., (*Q. grisea*, Liebm.)

Q. virens, Ait.—Live Oak.—Leaves two to four inches long, thick evergreen, oblong, obtuse, somewhat rough or wrinkled, smooth and shining above, hoary tomentose beneath, the margins revolute. Fruit on a rather long stem, cup top-shaped, wrinkled, enclosing the base of the oblong, brown acorn, which matures the second season. A large tree, with spreading branches, fifty to seventy feet high, and stem four to six feet in diameter. Wood yellowish or light brown, very heavy, fine-grained, and very durable. Formerly largely used for ship building, but less since iron has been employed for similar purposes. There are two or three varieties described in botanical works. A common and rather abundant tree in the Southern Atlantic States, and westward along the Gulf Coast to Mexico.

Q. Wislizeni.—A. DC.—Leaves smooth, dark green and shining, one to three, rarely four inches long, by one to two broad, varying in shape from narrowly-lanceotate to broadly oval, entire or serrate, or often sinuate, dentate or lobed. Cups turbinate, very deep, or even tubular, one half to an inch deep, covered with brown lanceolate scales. Acorns slender, tapering, often an inch and a half long. Leaves very persistent, and the acorns maturing in the second season. A large tree, fifty to sixty feet high, and common in the lower valleys of California. Var. *frutescens*, Engelmann, is a shrub from three to ten feet high, known in the Sierra Nevadas as Desert Oak.

HYBRID OAKS.

There have been from time to time single specimens of oak trees found in different parts of the country that did not appear to agree with the recognized distinctive characteristics of any of the indigenous species. It has been claimed for many years that some of the number at least were hybrids, and Dr. Engelmann favored this idea, and gives a list of those known to him in the monograph to which I have already referred. In this list he places the Bartram Oak, and the *Quercus Leana*, of Nuttall's Sylva, and several other unique forms of our indigenous oaks.

FOREIGN SPECIES AND VARIETIES.

Of all the foreign species of the oak, the European or English oak, *Quercus Robur* is probably the most familiar to the people of this country, and were we in want of any additional species for planting in forests, this one could be recommended, as it is closely related to our White Oaks, but we have such a large number of species of our own that we have no good reason for introducing anything from abroad of the kind, except for ornamental purposes. The European oak has yielded many beautiful varieties, among which I may name the Purple-leaved, Golden-leaved, Mottled-leaved, Cut-leaved, Weeping oak, and numerous varieties to be found named in nurserymen's catalogues. Besides these, there are several varieties of the Turkey Oak (*Q. cerris*) in cultivation, and recently several handsome species and varieties of the oak have been introduced from Japan, among them the noble Daimio Oak, which may at some future time be planted as a forest tree in this country, but my limited space will not admit of even enumerating the large number of species of foreign oaks, however much I might desire to do so,

RHAMNUS, Linn.—*Buckthorn.*

An extensive genus containing nearly sixty species of evergreen and deciduous shrubs or small trees. Flowers perfect, or
the sexes separated; petals four or five, but in some species,
entirely wanting. Fruit, berry-like, containing two to four
bony or horn-like, one-seeded nutlets. Of the six indigenous
species, three belong to the Pacific Coast, and three to the
Eastern States. Only two grow to a hight of twenty feet.

Rhamnus alnifolia.—L'.Her.—A small shrub, two to four feet
high, with ovate-oblong deciduous leaves. Flowers without
petals. Fruit black, New England to Washington Territory.

R. California, Esch.—California Buckthorn.—Leaves ovate-oblong to elliptical, one to four inches long, acute or obtuse,
mostly rounded at the base, slightly toothed or entire, evergreen. Petals very small; fruit, blackish-purple, with thin
pulp, a quarter of an inch in diameter, and two to three-seeded.
A spreading shrub, from five to eighteen feet high, throughout
California. Var. *tomentella,* is densely white, tomentose, especially on the underside of the leaves.

R. Caroliniana, Walt.—Carolina Buckthorn.—Leaves three to
four inches long, oblong, wavy, and finely serrulate on the margins, the slender petioles and many-flowered clusters pubescent; petals five, minute. Fruit round, three-seeded. A small
tree, sometimes over twenty feet high, but usually a low, much-
branched shrub. Long Island to Florida, and west to the
Rocky Mountains

R. crocea, Nutt.—Red-Berried Buckthorn.—Leaves evergreen,
thick, oblong or obovate, to orbicular, variable, an inch or an
inch and a half long. Flowers tetramerous, without petals.
Fruit one fourth of an inch long, two to four-seeded, bright red.
A branching shrub, four to fifteen feet high. Mountains and
hillsides of Southern California, and eastward into Arizona.
Berries eaten by the Indians, and said to color their veins red.

R. lanceolatus, Pursh.—Narrow-Leaved Buckthorn.—Leaves
oblong-lanceolate, acute. Flowers clustered, on short pedicels,
or scattered on longer pedicels. Seed black, as large as a grain
of pepper. A tall shrub, from Pennsylvania, southward to
Alabama, in swamps.

R. Purshiana, DC.—Bear Berry.—Leaves deciduous, two to
seven inches long, one to three wide, elliptic, mostly acute, obtuse at base, denticulate, somewhat pubescent underneath.

Flowers large, petals minute. Fruit black, broadly obovoid, a quarter of an inch long ; three-lobed and three-seeded. A shrub or tree, twenty feet high, the young branches downy. Mendocino County, California, northward to the British Boundary.

The common Buckthorn of Europe (*R. catharticus*) has been so long cultivated in this country for hedges, that it has run wild in many places, becoming a small tree with thorny branchlets, with ovate or oblong leaves, and fruit with three to four seeds.

RHIZOPHORA, Linn.—*Mangrove.*

Trees or shrubs of maritime swamps, with opposite entire, evergreen leaves. The branches throw out roots freely, which descend and take root in the mud, each branch being supported by its own roots, a single tree in this manner may extend over a large space. Only one species.

Rhizophora Mangle, Linn.—Mangrove.—Leaves obovate-oblong; peduncles two to three flowered. Flowers pale yellow, quite showy. Fruit, a small, one-seeded nut, which remains attached to the tree until it germinates. A small tree, in the maritime swamps of Southern Florida, Louisiana, Texas, and throughout Tropical America.

RHODODENDRON, Linn.—*Rose Bay.*

An extensive genus of several hundred species, widely distributed over the globe, mostly in cool or temperate climates. Principally evergreen trees and shrubs with showy flowers, usually in terminal umbels or corymbs. The *Rhododendron* and *Azaleas* are so nearly allied, in fact, scarcely distinguishable as a whole, that our modern botanists have classified them all under the one generic name of *Rhododendron*, separating them under sub-genera or in groups. They are all handsome, ornamental shrubs or small trees, and extensively cultivated among all civilized nations. We have about a half dozen species of the *Azalea* proper, and four or at most five of the *Rhododendrons*, but only one of the number grows tall enough to be classed among trees.

Rhododendron maximum, L.—Great Laurel Rose Bay.—Leaves obovate-oblong, abrupt acute, smooth and green on both sides. Flowers bell-shaped, white or pale rose color, spotted within with yellow or green, and usually about an inch broad. Usually a shrub from ten to twenty feet high, but in the mountains at the South sometimes a tree forty feet high. Wood very hard, but usually too crooked to be of much value except for fuel or making handles for small tools. More or less abundant along the banks of streams, from Canada to Florida, but keeping in high woods in the South.

The *R. Catawbiense*, Michx., has lilac-purple flowers, and is the parent of many of our most valuable cultivated varieties. Hybrids between this and several foreign species have been raised in great numbers. The California Rhododendron (*R. Californicum*, Hooker), is a low shrub, four to eight feet high, with leaves four to six inches long, and handsome rose-colored flowers, two inches in diameter. It is a very handsome and showy shrub, worthy of extended cultivation. The Lapland Rose Bay (*R. Lapponicum*, Wahl.), is a low dwarf or prostrate shrub, with leaves about a half inch long, and very small violet-purple flowers. Found in the mountains of Northern New York, and the New England States, also in Europe. The native species of *Azalea* are all shrubs, of only moderate size, and for this reason are omitted.

RHUS, Linn.—*Sumach.*

A large and widely distributed genus of more than a hundred and twenty species, some fourteen of which inhabit the United States. Leaves simple or pinnate. Flowers small, either perfect, or the two sexes separate on the same plant, or on separate plants ; usually greenish-white or yellowish, in axillary or terminal panicles or racemes. Fruit, a small, dry drupe, in branching open panicles or close, compact clusters or heads. The leaves of some of the species extensively employed in tanning certain kinds of leather. The resinous juice of one or more species in Japan, yields the well-known lacquer varnish of that country, while from the fruit, a peculiar and valuable vegetable wax is extracted. There are several species that are very poisonous to some persons, but not to others. Only two or three of our native species grow large enough to be classed as trees, but as each has some peculiar habit or properties that should be well known to the practical forester, I will refer briefly to all, and first to those known to be poisonous.

Rhus diversiloba, Torr. and Gray.—Poison Oak, Yeara.—Leaves composed of three ovate, obovate, or elliptical leaflets, one to three inches long, obtuse or acute, three-lobed or coarsely toothed. Flowers whitish, in loose axillary panicles. Fruit smooth and whitish. A plant with a slender stem, erect or climbing by rootlets, three to eight feet high. Very much resembling the *Rhus Toxicodendron* of the Atlantic States. Common from Southern California to British Columbia.

R. venenata, DC.—Poison Sumach, Poison Dogwood, Poison Elder.—Branchlets and leaf-stalks smooth; leaflets seven to thirteen, ovate or oblong, abruptly pointed. Fruit small, globular, dun color, in loose axillary panicles, hanging on late in winter. A rather handsome, upright shrub or tree, sometimes twenty feet high. In swamps and low ground. Supposed to be the most poisonous of all the species, but there are many persons who can handle it with impunity.

R. Toxicodendron, Linn.—Poison Ivy, Poison Oak.—Leaves composed of three rhombic-ovate leaflets, mostly pointed and rather downy beneath, variously cut-lobed or toothed. Fruit same as the last, but leaves usually yellow after frosts, but sometimes slightly tinged with red. Usually climbing by rootlets, over rocks or ascending trees to a great hight, and the stem becoming as large as a man's arm. A species quite variable in form of growth, but always readily distinguished by its leaves and fruit. Michaux describes a low growing, southern form, under the name of the Oak-leaved (*quercifolium*), of a more erect habit, with variously lobed leaves, but the leaflets are only three in number. A common plant throughout the Atlantic States and westward to the Rocky Mountains. The following are all innoxious species, and some of them cultivated for ornament.

R. cotonoides, Nutt.—American Cotinus.—Leaves simple, thin, oval-obtuse, entire, acute at the base, the upper ones long petioled. Flowers perfect in an open panicle, the pedicles mostly abortive, elongating, and becoming plumose as in the common Smoke-tree or Venetian sumach tree in gardens. Nuttall says that during his tour into the interior of Arkansas Territory, in 1819, he discovered this species on the high, broken calcareous rocky banks of the Grand River, near a place called "Eagle's Nest." A large shrub, but it has recently been reported to have been found growing in Alabama, to the hight of twenty feet or more, with a stem nearly or quite a foot in diameter. Not poisonous, neither are any of the following species except the last.

R. typhina Linn,—Staghorn Sumach.—Leaflets, eleven to thirty-one, lanceolate-pointed, serrate, smooth, pale beneath. Young branches, leaf stalks, and fruit, densely velvety or hairy. Fruit red, acid to the taste, and in a dense, close, upright, terminal panicle. A common, low shrub, but sometimes a tree twenty feet high, with orange-colored, brittle wood. Often a great nuisance, appearing in neglected fields, and throwing up suckers from the large, coarse, subterraneous stems.

R. glabra, Linn.—Smooth Sumach.—Branches and leaves smooth, not downy. Leaflets eleven to thirty-one, whitish underneath, lanceolate-oblong, pointed and serrate. Fruit red, but in an open and spreading cluster. A small or large shrub, sometimes ten or twelve feet high, and common in rocky soils. Var. *laciniata*, or the Cut-leaved sumach of gardens, belongs to this species, and was found in Pennsylvania by the late Dr. Darlington, nearly a half century ago.

R. copallina, Linn. — Dwarf Sumach. —Young stalks and branches downy, petioles winged or broadly margined between the nine to twenty-one, oblong or ovate-lanceolate leaflets, mostly entire, smooth above and downy beneath. Fruit red. A low shrub, four to eight feet high, on the borders of woods, in both Northern and Southern States.

R. aromatica, Ait.—Fragrant Sumach.—Leaves composed of three cut-lobed leaflets, of a rhombic-ovate form, downy when young, aromatic-scented. Flowers light yellow, and appear in spring, before the leaves. Fruit red, in short spikes. Var. *trilobata*, Gray, or *R. trilobata*, Nutt., is found from Texas to Washington Territory. Fruit pleasant tasted, and eaten by the Indians. The small, slender twigs, are also employed for making very choice baskets.

R. pumila, Michx.—Dwarf Sumach.—A low growing shrub, with eleven to thirteen oval, oblong, pointed leaflets, coarsely serrate, and downy beneath. Fruit red and hairy. In Georgia and North Carolina.

R. integrifolia, Benth. and Hook.—Entire-leaved Sumach.— Leaves evergreen, pubescent when young, but soon smooth ; broadly ovate, usually entire, but sometimes spiny-toothed ; one to three inches long. Flowers rose-colored, in close panicles, one to three inches long. Fruit dark-red, viscid, ovate, nearly a half inch long. Nuttall, in describing this species, under the name of *Styphonia integrifolia*, says that it is an unsightly tree, with a stem about the thickness of a man's arm, branching in a wide and straggling manner, forming impervious thickets along the margins of cliffs and steep banks near

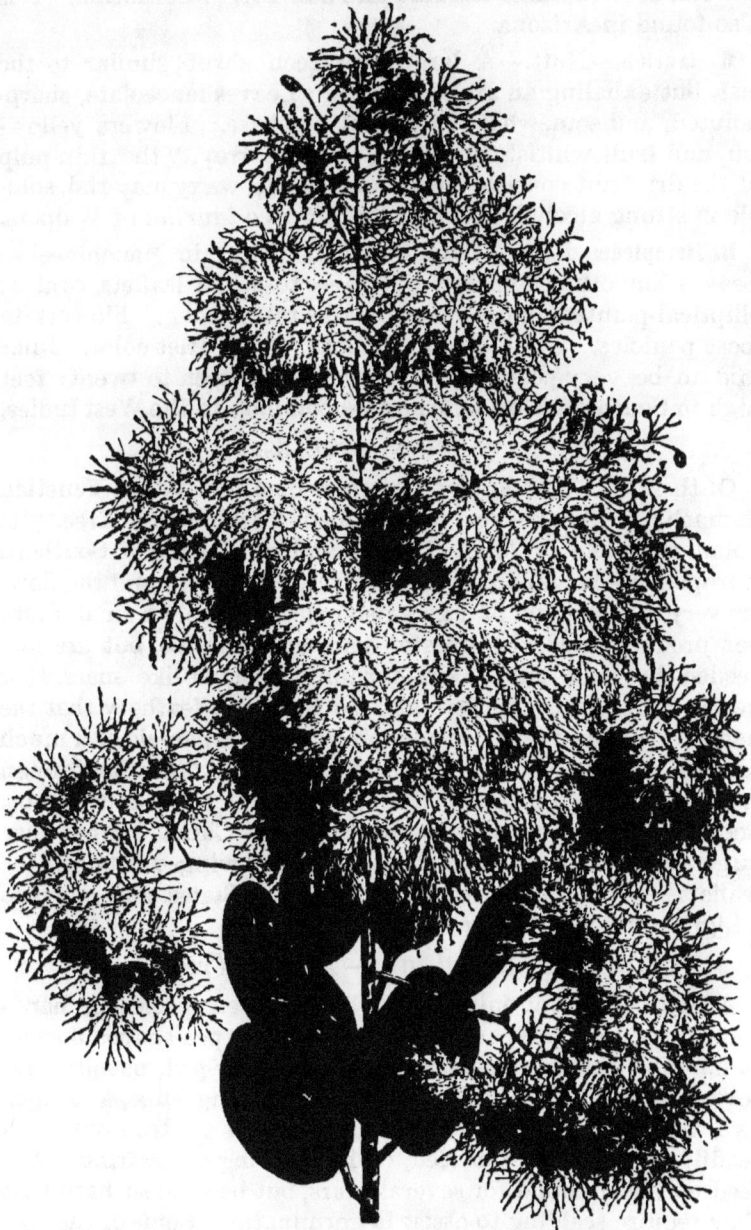

Fig. 47.—VENETIAN SUMACH, OR SMOKE TREE.

the sea, around Santa Barbara and San Diego, California. It is also found in Arizona.

R. laurina.—Nutt.—A large evergreen shrub, similar to the last, but exhaling an aromatic odor. Leaves lanceolate, sharp-pointed, and somewhat rounded at the base. Flowers yellowish, and fruit whitish, according to Dr. Torrey, "the thin pulp of the dry fruit consists chiefly of a white, waxy material, soluble in strong alcohol." This is the *Lithræa laurina* of Walpers.

R. Metopicum, Linn.—Coral Sumach, Mountain Manchineel.— Leaves smooth, composed of three to seven leaflets, oval or elliptical-pointed, entire. Leaf-stalks rather long. Flowers in loose panicles. Fruit oblong, smooth, of a scarlet color. Juice said to be very poisonous. A rare tree, fifteen to twenty feet high in Southern Florida, but more common in the West Indies.

FOREIGN SPECIES.

Of these, the best known and most common is the Venetian Sumach, or Smoke-tree, also called Purple-fringe tree. Its botanical name is *Rhus Cotinus,* and is a native of Southern Europe. Its leaves are roundish-oval, or oblong, and the flowers very minute, and of a greenish color, but only a small number produce seed, the greater part are abortive, but are succeeded by long, silky hairs, forming a cloud-like mass, that nearly conceals the foliage, and so light and feathery that the name of smoke-tree is not unappropriate. A panicle of a much reduced flower-cluster is shown in fig. 47. A few other foreign species have been introduced as ornamental trees and shrubs, one of the best of these is *R. Osbecki,* from China. It has very large pinnate leaves, the leaf-stalk broadly winged between the leaflets. It is quite hardy in our Northern States, and grows to a hight of twenty feet or more.

ROBINIA, Linn.—Locust Tree.

A small genus of only about a half a dozen species, of handsome deciduous trees or shrubs, with showy, pea-shaped flowers in hanging axillary racemes. Fruit, a linear-pod, usually flat, several seeded, margined on the seed-bearing edge, at length two-valved, opening and allowing the seed to drop out. All readily propagated from seed, or by budding or grafting. The seed will keep sound for several years, but become so hard that they require scalding to assist in germination. Some of the species have rather strong spines on the smaller branches, others only armed with slender prickles.

Robinia Pseudacacia, L.—Common Locust or False Acacia.— Leaves composed of from nine to seventeen small, oblong-ovate leaflets. Flowers white, fragrant, in pendulous racemes, three to five inches long; pods flat, containing four to six hard, small, and rather flattish seeds. Usually a slender tree, sixty to eighty feet high, with stem two to three feet in diameter. Wood white, or greenish-yellow, very hard and close-grained, and when of slow growth, it is very durable, but when grown on very rich soils, as for instance on the rich western prairie soils, it is far less durable than when raised on lighter and poorer land. When planted singly or in small groves, the trees are usually infested with borers, but in larger plantations or forests the insect confines itself mainly to trees, the stems of which are exposed to the direct rays of the sun. A well known tree, extensively naturalized in all of the Atlantic States, but native of Southern Pennsylvania and southward along the mountains, and by some authors said to extend west to Missouri. It is a tree very much inclined to spread by suckers from the roots, as well as from the seed, which are usually widely scattered by winds.

R. viscosa, Vent.—Clammy Locust.—Small branches and leafstalks clammy, spines very small. Leaflets eleven to twenty-five, ovate and oblong, obtuse or slightly heart-shaped at base, slightly downy beneath, tipped with a short bristle. Flowers in a short, rather compact, roundish, upright raceme, rose-color and inodorous. Pods three to five-seeded. A small tree, from thirty to forty feet high. Wood said to be valuable. Native of North Carolina and Georgia in the mountains, along the banks of streams. Often cultivated as an ornamental tree. Produces suckers in great abundance if the roots are disturbed or broken.

R. hispida, Linn.—Bristly or Rose Acacia.—Branches thickly covered with small, slender bristles. Leaflets eleven to eighteen, ovate, or oblong-ovate, rounded at the base, and tipped with a long bristle. Flowers large, in loose, and mostly pendulous racemes; bright pink or rose-color, very showy and handsome. There are several wild and cultivated varieties, all low, straggling shrubs, their roots running in light soils to a great distance, and producing numerous suckers.

SALIX, Tour.— *Willow, Osier.*

An immense genus of about a hundred and sixty species, the larger part belonging to Europe and Asia, a half dozen inhabiting Africa and South America, and about sixty species in North America. A very difficult genus to elaborate, as all the species are more or less variable. Staminate and pistillate aments or catkins preceding or accompanying the leaves. Seeds minute, without albumen, cotyledons flattened. Trees, shrubs, or low undershrubs, with alternate simple leaves. Wood light, soft, of little economic value. The bark containing a bitter principle, known as *salacin*, sometimes used as a substitute for quinine. The slender, tough twigs, of some of the species are extensively used for basket-making, and cultivated for this purpose, a few species for ornament, and a far less number for their wood. As a whole, the willows are of no great economic importance, but all are readily propagated by cuttings, and some of the smaller species are of value for planting in drifting sands, and the banks of streams for the purpose of holding the loose soil in place. My limited space will not admit of enumerating all the indigenous and unimportant species, therefore I will only name a few of the larger-growing native, and foreign species and varieties.

Salix cordata, Muhl.—Heart-leaved Willow.—Leaves oblong-lanceolate, taper-pointed, truncate or heart-shaped at base ; sharply toothed, smooth above, pale, downy beneath ; catkins appearing with the leaves, leafy at base, cylindrical, the fertile ones elongating with the development of the seeds. A small tree, sometimes twenty feet high. From the North Eastern States to the Arctic Coast. Abundant in Colorado, Utah, and Nevada. There are some four or five local varieties recognized by botanists.

S. lævigata, Bebb.—Smooth-leaved Willow.—Leaves lanceolate, or oblong-lanceolate, sharp-pointed, three to seven inches long, smooth, glossy-green above, whitish beneath, minutely serrulate, the male catkins, roundish-obovate, the female narrower, and truncate, with two to four irregular teeth at apex. An erect, pyramidal tree, fifteen to fifty feet high, with a stem one to two feet in diameter, with fissured, dark-brown bark. In California, along the bottom lands near streams from San Diego County to the Sacramento Valley. There are several varieties.

S. lasiandra, Benth.—Long-leaved Willow.—Leaves lanceolate, taper-pointed, roundish at base, smooth, whitish beneath, margins closely and sharply serrate ; catkins with a leafy stalk, with thin, yellowish scales, more or less hairy at the base ; the female catkins smooth. Three natural varieties are described by Dr. Bebb, in Botany of California, viz. Var. *typica*, var. *lancifolia*, and var. *Fendleriana*. A tree twenty to sixty feet high. Sacramento Valley, California, and northward to British Columbia. Some of the varieties extend eastward to New Mexico.

S. lasiolepis.—Benth.—Leaves oblanceolate, or rarely oblong-lanceolate, four to six inches long, one half to one inch wide, the lower ones spatulate, more or less pubescent, especially when young ; catkins sessile, one to three inches long, cylindrical, densely flowered, stamens yellow, three times as long as the scales. Var. *Bigelovii*, Bebb., has leaves more obovate than the species, and var. *fallax*, Bebb. Leaves lanceolate-oblong, abruptly contracted at base. A large tree, forty to sixty feet high in the neighborhood of San Francisco, and southward in California. Said to be a common tree throughout the State.

C. lucida, Muhl.—Shining Willow.—Leaves ovate-oblong, or lanceolate, with a long, tapering point, smooth and shining on both sides, serrate. A handsome species of willow, rarely more than twenty feet high, along the banks of streams from Pennsylvania, northward through British America.

S. nigra, Marshall.—Black Willow.—Leaves narrowly-lanceolate, pointed and tapering at each end, serrate, smooth, except on the petioles and midrib, bright green on both sides. A small tree, twenty to forty feet high, with a rough, black bark, hence the specific name. A rare tree in Northern New England and Canada, but more common south and west, extending entirely across the continent, being plentiful in California.

FOREIGN SPECIES AND VARIETIES.

Among the foreign species of the willow, there are quite a number that are better known and far more common in cultivation than any of our native species, in fact, it may be said that our indigenous willows are almost unknown among cultivated trees and shrubs, while several of the foreign ones may be seen in almost every garden, park, or pleasure ground in the country. The old and familiar Weeping Willow must have been introduced at a very early period in our history, for very old and very large trees of it may be seen in all of our older States, and specimens with stems four to six feet in diameter, near the ground, are far from being rare or uncommon, and while this tree is of little or no economic value, it has long been a favorite ornamental tree for planting about ponds, churches, and in

10

cemeteries, and will probably continue to have its admirers for all time. It is sometimes called the Babylonian Willow, but it is certainly not a native of Babylon or any other hot climate, but is without doubt a native of Northern Asia.

The Hoop-leaved or Ring-leaved Willow (*S. annularis*), is a variety of the so-called Babylonian Willow, with leaves curved into a ring, but of a similar weeping habit. The next most familiar species is the White Willow (*S. alba*), which has been highly extolled, and quite extensively planted for fencing and fuel in the Western States, and while it is a rapid growing tree, and the wood moderately firm and good for a willow, still it is an inferior forest tree, and scarcely worth cultivating where other and better species will grow. This species, and its variety with yellow twigs (*S. alba*, var. *vitellina*), are quite common along the banks of small streams and ponds in the Eastern States, where they have been planted for ornament, or shade for stock in pastures. All of these large growing willows have very large masses of fine fibrous roots that penetrate the soil to a great depth, and will push to a great distance in search of moisture, and for this reason they should never be planted near drains, wells, or where their roots will be likely to do injury to these or similar structures. There are many small ornamental varieties that are well worthy of a a place in gardens and pleasure grounds, but as they are fully described in nurserymen's catalogues, and are of no considerable economic value, I omit further reference to them here.

SAMBUCUS, Tour.—*Elder*.

A small genus of no especial importance, although some of the species have had some reputation for their medicinal properties. They are principally small shrubs, with one European and one Asiatic herbaceous species. Flowers small but numerous, in compound cymes or clusters. Fruit, a small, round, juicy drupe, but usually called a berry, containing several separate seed-like nutlets, each with one seed. Only one species that becomes a tree.

Sambucus glauca, Nutt.—Tree-Elder.—Leaflets three to nine, of firm texture, ovate or lanceolate, sharply serrate, with rigid, spreading teeth. Flowers in a broad, flat cluster or cyme. Fruit black, but with so much bloom, that they appear to be white; pith of shoots white. A small tree, but sometimes twenty feet high, and a stem a foot in diameter. Wood like that of all the elders, very hard, but owing to large pith, only valuable for a few purposes. Common in California and southward, also in Oregon and Washington Territory, also in the valleys throughout the Rocky Mountain regions. The European Elder (*S. racemosa*) is also common in the Rocky Mountains, and eastward in our more Northern States, in high, rocky situations, but the Eastern form is known as var. *pubens*, Michx. Our most common species is the Black-Berried Elder (*S. Canadensis*), the fruit of which is extensively used for making a kind of wine or cordial. There are several ornamental varieties in cultivation, one with golden variegated leaves, another with silver variegated, also a cut-leaved, and one with double white flowers.

SAPINDUS, Linn.—*Soap-Berry.*

A genus of a dozen or more species of evergreen trees, principally tropical, noted for the saponaceous properties of the pulp (*aril*), surrounding the seeds of some of the species. This substance is used as a substitute for soap in South America, and is said to lather quite freely in water. The flowers are produced in axillary or terminal racemes or panicles. Leaves abruptly pinnate. Seeds horny. Two species inhabit our southern borders.

Sapindus marginatus, Willd.—Soap-Berry.—Leaflets nine to eighteen, opposite or alternate, ovate-lanceolate, unequal-sided, strongly veined above; panicles large, and dense flowered. Flowers white. Fruit globose. A tree from twenty to forty feet high from Georgia to Florida, and near the Coast westward to Southern Arizona, also in Mexico.

S. Saponaria, L.—Soap-Berry.—A common tree in the West Indies, and said to be found sparingly in Southern Florida. The fruit known as Indian soap, they are as large as cherries, and the nut-like seed shining black, and were formally much used in England for buttons, sometimes being tipped with silver.

SASSAFRAS, Nees.—*Sassafras.*

A genus of the *Lauraceæ* or Laurel Family, and still classed in many botanical works under the generic name of *Laurus*. A well known tree with small greenish or whitish flowers in clustered racemes, appearing before the leaves. We have only one species, the

Sassafras officinale, Nees.—Sassafras.—Leaves deciduous, ovate entire, or two to three-lobed, smooth or pubescent, exceedingly variable in size and shape on the same tree. Fruit blue, on thick red petioles. The pulp thin, but of a succulent, rather spicy flavor, greedily eaten when ripe by birds. The bark, and especially that on the roots strongly aromatic, and formerly was in high repute as a medicine for various diseases. It is still in some demand, as the bark of the roots is a powerful stimulant; but the oil distilled from the roots is now of more commercial importance, it being extensively used in imparting flavor to candies and similar articles. A rather handsome tree, fifty to sixty feet high in favorable soils, with a stem two feet in diameter. Wood reddish in very old trees, moderately hard, easily worked, and considered very durable. Bark on young twigs very smooth, and of a deep green color, but on older branches and stems rough, of a grayish color, and deeply furrowed. The roots produce suckers in great abundance, and these are not readily destroyed as their roots penetrate the soil to a great depth. Common on light soils, river banks, and in rocky woods, from Canada to Florida, and west to Texas.

SCHÆFFERIA, Jacq.—*Crab Wood.*

A genus of the *Celastraceœ* or Staff-tree Family, with evergreen alternate leaves, and diœcious flowers; very small, greenish, and in axillary clusters. The one species found in the United States has been described under three different names, viz. *S. completa*, Swartz, and *S. buxifolia*, Nutt., but now recognized as the

Schæfferia frutescens, Jacq.—Crab Wood, Jamaica Boxwood.— Leaves obovate-oblong, entire, acute or obtuse, an inch and a half long. Flowers three to five in a cluster, the slender stalks arising from a wart-like peduncle. Fruit a two-celled, two-seeded drupe. A small tree, with a hard, coarse-grained wood. Southern Florida, and in the West Indies.

A closely allied tree, the *Schœpfia arborescens*, R. and S., inhabiting the West Indies, is reported to have been found in Southern Florida. It is, however, a very small tree of not much importance, although interesting to the botanist.

SEBASTIANIA, Muell.—*Poison Wood.*

Tropical or sub-tropical trees and shrubs, with milky juice; alternate, serrate or crenate leaves. Flowers diœcious or monœcious without petals.

Sebastiania lucida, Muell.—Shining-Leaved Poison Wood.—

Leaves smooth, coriacious, obovate-oblong, obtuse or emarginate, crenate. Fertile flowers, solitary or in pairs, with long stems ; sterile ones very minute in cylindrical spikes. Fruit, a capsule of three, one-celled, one-seeded, two-valved carpels. Described under the generic name of Excœcaria in Chapman's Flora of the South, Nuttall's Sylva, etc. A tree thirty to forty feet high, with yellowish-white, hard, and close-grained wood. In South Florida.

SHEPHERDIA, Nutt.—*Rabbit Berry*.

A genus of only three species, all found in the United States, and in the more northern regions. They are small trees or shrubs, with diœcious flowers, the sterile with a four-pointed calyx, the fertile, with an urn-shaped, four-cleft calyx ; the fertile flowers much the smallest. Fruit small, red, yellow or scurfy. Only one species that grows to a hight of twenty feet.

Shepherdia argentea, Nutt.—Buffalo Berry, Rabbit Berry — Leaves oblong, ovate, silvery on both sides ; male flowers in clusters, the calyx yellow inside, but silvery on the outside ; female flowers very minute, scarcely noticeable on the plant without close inspection, as they are of a dull gray color. Fruit collected into clusters, sometimes in such abundance as to entirely surround the smaller branches, bright scarlet, resembling small currants ; juicy sub-acid and pleasant flavored. Excellent jelly is made from the fruit, and some persons think it superior to currant jelly. A handsome small tree, with grayish rough bark and hard wood. The branchlets are terminated by a sharp thorn, and for this, and other reasons, this species has been recommended for hedges in cold northern localities. As the two sexes of flowers are on different plants, it is necessary to at least have one of each growing near together in order to obtain fruit, and I do not know of any way of determining the sex of the plants except by waiting until they come into bloom, then each should be labelled if they are to be transplanted or set out for fruit, but one staminate will suffice for a half dozen pistillate plants, if set near together. The trees are readily propagated from seed. Native of Northern New Mexico and through the Rocky Mountain regions, northward to British America. The two other native species are low shrubs, *S. Canadensis*, Nutt., is a low, scurfy shrub, with ovate leaves, and yellowish-red insipid fruit, found from Vermont, westward to the Pacific Ocean. *S. rotundifolia*, Parry, with small, crowded leaves, and scurfy fruit, is a low shrub, peculiar to the mountains of Southern Utah.

SIDEROXYLON, Linn.—*Ironwood.*

Tropical trees and shrubs of the Sapodilla Family. Flowers with a four-pointed calyx and corolla, five-cleft. Fruit, a small drupe, mostly one-celled and one-seeded, the kernel with abundant albumen.

Sideroxylon Mastichodendron, Jacq.—Mastic Tree.—Leaves smooth, five to six inches long, very thin, elliptical, obtuse, wavy on the margins; on slender petioles. Flowers few in a cluster and small. Fruit purplish, ovoid. A large tree in the West Indies, but only thirty to forty feet high at Key West, Florida.

SIMARUBA, Aublet.—*Quassia.*

Trees or shrubs with bitter, milky juice, pinnate-alternate leaves, and small greenish monœcious or diœcious flowers. Fruit drupaceous and one-seeded. There are several tropical species, but only one coming within the United States.

Simaruba glauca, DC.—Bitter Wood.—-Leaves and twigs smooth. Flowers diœcious ; leaflets four to eight, alternate and opposite, coriaceous, obovate or oblong, obtuse, paler beneath than above. Fruit oval, mostly solitary. According to some authorities, a small tree or shrub. Chapman says, "a large tree in South Florida."

SOPHORA.—Linn.

A genus of trees, shrubs, and herbaceous plants, and of about twenty-five species, mainly in the warmer parts of the world, although a few inhabit the colder regions of Asia. Leaves unevenly pinnate, with few or many entire and sometimes quite thick leaflets. Flowers showy, pea-shaped, succeeded by large, thick pods, with several seeds. We have no indigenous species worthy of any especial attention, and only one grows to the hight of twenty feet, and this is the

Sophora secundiflora.—Lag.—Leaves evergreen. Flowers blue and quite showy ; sweet scented. A small tree, twenty to thirty feet high, with very hard, heavy yellow wood, said to be excellent for fuel. In groves near Matagorda Bay to Western Texas. The next largest native species is *S. affinis,* Torr. and Gray. A large shrub, ten to fifteen feet feet high, with evergreen leaves, and very hard wood. In Arkansas and Eastern Texas. *S. Ari-*

zonica, Watson, is a low, evergreen shrub, found in Western Arizona. The *S. tomentosa*, Linn., is a small evergreen shrub, four to six feet high, in Florida, along the coast. The Japan Sophora (*S. Japonica*) is a well-known ornamental tree, introduced many years ago. The weeping variety has long been a favorite tree for planting on lawns, as it is one of the

Fig. 48.—FLOWER OF STUARTIA PENTAGYNIA.

most picturesque and graceful of the pendulous-branched trees, and quite hardy in most of our Northern States.

STUARTIA, Catesby.—*Stuartia.*

A genus of only a few species of deciduous shrubs or small trees, belonging to the Camellia Family. Flowers large and showy, and highly prized ornamental plants. Fruit a five-celled pod, and with one to two small seeds in each. We have

two native species, and one in cultivation, recently introduced
from Japan.

Stuartia pentagynia.—L.'Her.—Leaves oval-acute, finely pubescent, serrate. Flowers with five large crimped-edged petals,
with purple spot at the base. Flowers appear late in spring or
early summer, and of the size and form shown in fig. 48. A
large shrub or small tree, from the mountains of North Carolina and Georgia. The wood very hard, white. The plants are
hardy as far north as New York City, and in my grounds in
Northern New Jersey.

S. Virginica.—Cav.—Leaves oval, thin serrulate, finely pubescent. Flowers two to three inches broad, with purple stamens.
A large shrub or small tree, not hardy north of Washington,
and found in shady woods from North Carolina to Florida.

S. Japonica is a rare shrub, introduced a few years ago from
Japan by the Parsons & Sons Company, Flushing, N. Y.
Flowers small, with yellow stamens. A variety of this, *S. Japonica grandiflora*, has larger flowers than the species, both
handsome and desirable ornamental shrubs.

SWIETENIA, Linn.—*Mahogany.*

A genus of a few species of large evergreen tropical trees,
highly valued for their excellent wood. Flowers small, green,
or reddish-yellow, in spreading axillary panicles. Fruit a large
five-celled and five-valved capsule containing seeds, imbricated
in two rows.

Swietenia Mahogoni, Linn.—Mahogany Tree.—Leaves alternate,
abruptly pinnate, and composed of six to ten opposite, entire,
ovate-lanceolate leaflets. Flowers greenish-yellow, three-eighths
of an inch broad. A large and rare tree in South Florida, but
formerly very abundant in the West Indies, but now becoming
scarce, owing to the great demand for the wood, which is extensively used for all kinds of cabinet work.

SYMPLOCOS, Jacq.—*Sweet Leaf.*

A genus of about a half dozen species of small trees or
shrubs, with evergreen or very persistent leaves, and small but
showy flowers. The species are widely distributed in China,
Japan, and in Mexico, and one in our Southern States, the last is

Symplocos tinctoria, L.'Her.—Sweet Leaf, Horse Sugar.—Leaves
simple, smooth, oblong, rather persistent, almost evergreen.
Flowers yellow, six to twelve in a sessile cluster. Fruit a one-

seeded berry. Leaves have a sweetish taste, and are greedily devoured by cattle. A small tree with rather firm wood. Delaware and South to Florida, and westward to Louisiana.

TILIA, Linn.—*Basswood, Linden.*

A genus of only about a half dozen species of deciduous trees, inhabiting the temperate regions of Europe and America. They are all handsome and valuable trees, with soft and white wood. Leaves more or less heart-shaped ; often soft and downy. Flowers with five spatulate, oblong petals, cream-color and in small cymes or clusters, hanging on an axillary slender peduncle, which is attached to a long, slender, and thin leaf-like bract. Fruit a small globular nut, one-celled, and one or two-seeded. The flowers are fragrant and yield a large quantity of clear white delicate-flavored honey. We have but two indigenous species.

Tilia Americana, Linn.—American Basswood.—Leaves four to six inches broad, smooth, and green on both sides, obliquely heart-shaped, sharply serrate. Bark and buds mucilaginous, the inner bark used for making bass-mats, also for coarse cordage, and employed by nurserymen for tying in buds in the propagation of fruit, and other trees and plants. To prepare the bark for use it is stripped from the trees in spring, and placed in water until the mucilaginous properties have been dissolved, then taken out, and divided into thin layers. Wood white, soft, and light, easily worked, and extensively used for inside work of various kinds. A handsome large tree, sixty to eighty feet high, with stem two to four feet in diameter. Common in moist soils, and along streams in all of the States east of the Rocky Mountains, except in the extreme Southern. It is found in the mountains of Georgia, and northward to Canada and Lake Superior.

Var. *pubescens*, Gray, has rather thin leaves, softly pubescent underneath. A smaller tree than the species, and is found in the swamps of North Carolina and Florida, near the coast.

Var. *macrophylla* is a cultivated variety, with leaves much larger than those of the species, and is a very handsome, rapid growing tree.

T. heterophylla, Vent.—White Basswood.—Leaves large, often eight inches broad, smooth and bright green above, silvery-white, and downy underneath. A handsome tree, thirty to fifty feet high, in the mountains of Pennsylvania, Kentucky, and southward to North Carolina and Georgia.

FOREIGN SPECIES AND VARIETIES.

Some of the European botanists make several species of the different forms or varieties of the *Tilia Europœa*, while others consider them as all belonging to one. For all practical purposes they may be considered as one with many well marked and interesting varieties. The most common European Linden has large, deep green leaves, but distinguished from the American species by the absence of the petal-like scales among the stamens in the flowers. The tree grows to a very large size, and many of them from fifty to a hundred years old, may be found in and about our Eastern cities and villages. The wood is similar to that of the American Basswood, but the twigs are more numerous and slender, and the tree is of a more compact habit than that of the *T. Americana*. The following are some of the most common and desirable varieties of the European species, (*T. Europœa*).

Var. *alba* or *argentea*, a very graceful, vigorous-growing tree, with whitish leaves.—Var. *alba pendula*, a weeping variety of the last, with slender twigs and whitish leaves.—Var. *alba pendula variegata*, a weeping variety with variegated leaves.— Var. *aurea*, the bark on the twigs of a bright golden color in winter.—Var. *aurea platiphylla*, twigs yellow, but leaves longer than in the above.—Var. *dasystyla*, a variety or species from Tauria, with large, heart-shaped dark-green leaves, and yellow twigs. Some authorities consider this a distinct species.—Var. *laciniata*, leaves finely divided, or cut and twisted.— Var. *laciniata rubra*, leaves finely cut, and the bark on young twigs red in winter.—Var. *platiphylla*, similar to the species, but with somewhat larger leaves.—Var. *pyramidalis*, a tree of rapid growth and pyramidal habit, and reddish bark on the twigs.—Var. *rubra*, the common red-twigged European Linden.—Var. *vitifolia*, leaves resembling those of the grape vine ; bark on young twigs bright red.

ULMUS, Linn.—*Elm*.

A genus of less than twenty species, but an immense number of wild and cultivated varieties, principally native of North America and Europe. They are mostly lofty, deciduous trees of rapid growth. Flowers usually perfect, yellowish or purplish in lateral clusters ; in our species appearing in spring, before the leaves. Fruit a one-celled and one-seeded membraneous samara, winged all round. Propagated by seeds, layers, or budding and grafting. All the species produce a large number of tough, fibrous roots, not at all difficult to make grow when transplanted.

Ulmus alata, Michx.—Winged Elm, Small-leaved Elm, Wahoo.—Leaves ovate-oblong, one inch to an inch and a half long, sharply serrate, acute, commonly rounded at the base, rough above and downy beneath. Flowers clustered on slender stalks. Fruit oval and downy on the margins. A small tree, thirty to forty feet high, with corky branches and hard, compact wood, very difficult to split, and for this reason extensively employed for the hubs of wagon and carriage wheels. In Virginia, Kentucky, and southward, also west of the Mississippi, in Nebraska, and south to Texas.

U. Americana, Willd.—American Elm.—Small branches and twigs smooth. Leaves three to four inches long, thin, obovate-oblong or oval, abruptly pointed, sharply serrate, rather soft and velvety beneath. Flowers in close clusters or bundles. Fruit smooth, except on the margins. A well known, very large and common tree, with a wide spreading head and long, slender, drooping branches. Wood brown, very tough in young trees, light and moderately strong in the old, always difficult to split, extensively used in the manufacture of hubs, small trees being used for this purpose. The wood of large trees used for planks, chair seats, and various other purposes where it is not exposed to the weather. Common in moist soils throughout the United States and Canada, east of the Rocky Mountains.

U. crassifolia, Nutt.—Small-leaved Elm, Opaque-leaved Elm.—Leaves small, only about an inch long, oblong-oval, rough, serrate. Fruit small, oval or elliptic, rather deeply divided or toothed at the summit. Young branches smooth, thickly studded with leaves. A curious and handsome little tree in Southwestern Arkansas to Southern Texas. Figured and described in Nuttall's North American Sylva, Vol. 4, under the name of *U. opaca.*

U. fulva, Michx.—Slippery Elm, Red Elm.—Buds in spring soft and downy, with large rusty hairs, branches also somewhat downy. Leaves thick, ovate-oblong, taper-pointed, doubly serrate, very rough above, velvety beneath, often six to eight inches long. Inner bark very mucilaginous, and extensively employed as a medicine. A medium sized tree, with red-

dish hard wood, more durable than that of any other indigenous species, and is often used for fence posts and rails.

U. racemosa, Thomas.—Corky White Elm.—Twigs and bud
scales downy, and branches often with corky ridges. Leaves
obovate-oblong, oblique at the base, sharply serrate. Flowers
in racemes, and not in compact or close clusters, as in our
other native species. Fruit smooth, oval or elliptic, very downy,
with the margins thickly fringed. A large tree, with fine-
grained, heavy, tough wood, superior to that of any other indigenous elm for purposes where toughness and elasticity is
required. Common in Western New York, Ontario, and west
to Nebraska. Also said to be found in Ohio, Michigan and
Kentucky.

FOREIGN SPECIES AND VARIETIES.

The elms are all so inclined to vary from seed, that it is often
difficult to determine the original type of a species, as well as to
determine to what species many of the cultivated varieties belong. What is called the English Elm (*Ulmus campestris*), was
early introduced into this country, and quite extensively planted
in and about Boston, where at this time some of the most noble
specimens of this tree can now be seen in full maturity, and of
great age. The leaves of this species differ from our common
American Elms, in being smaller and of a darker green color,
and the branches spring out from the main stem more nearly at
right angles, and the general form of the head is more inclined
to be pyramidal than broadest at the top, as usual with our
American White Elm, which has been so extensively planted in
nearly all New England cities and villages. The English Cork-
bark Elm (*Ulmus suberosa*) is a variety of the last, with its
young branches very corky, with rougher and much larger
leaves, while the variety known as the Dutch Cork-bark (*U.
major*), has still larger leaves, and of a more spreading habit of
growth. Some of the largest specimens of both the English and
Dutch Cork-bark Elms to be found in this country, are to be
seen on Long Island, near Glen Cove and eastward, some of
them probably are two centuries old.

The Scotch or Wych Elm (*U. montana*) is a noble tree, with
wide spreading branches, and although not so well known in
this country as the English or Dutch Cork-bark, still of late
years it has been more largely planted than formerly. There is
a weeping variety, the (*U. m. pendula*), and a smooth-leaved

one (*U. m. glabra*), also another known as the Exeter Elm (*U. m. fastigiata*), noted for its peculiar fastigiate growth and twisted leaves. There are also, in addition to the varieties named, at least a score of others of similar origin, described in nurserymen's catalogues, all desirable for ornamental purposes.

UMBELLULARIA, Nutt.—*Spice Tree.*

A genus closely allied to the Sassafras Tree of our Eastern States, but found only on our West Coast. Flowers yellowish-green; perfect in pedunculate umbels, which are enclosed before expansion in a four-bracted involucre. A tree with thick, fragrant, evergreen leaves. Only one species.

Umbellularia Californica, Nutt.—Spice Tree, Mountain Laurel.— Leaves green and shining, lanceolate-oblong, acute at each end, or sometimes rounded at base, two to four inches long, short petioled. Fruit an ovate-elliptical or globose drupe, nearly an inch long, dark purple, with thin pulp and stone. The foliage is exceedingly acrid, exhaling a pungent odor, which excites sneezing. This tree is also known as the California Olive, California Laurel, Cajeput, etc., etc. The fruit is very persistent, remaining on the trees all winter. A small tree in the Coast Ranges of California, but extending northward to Oregon, where it grows to a hight of nearly or quite one hundred feet, and a stem two to four feet in diameter. Wood of a brownish color, close-grained, handsome and valuable, much used for cabinet work and wainscoting.

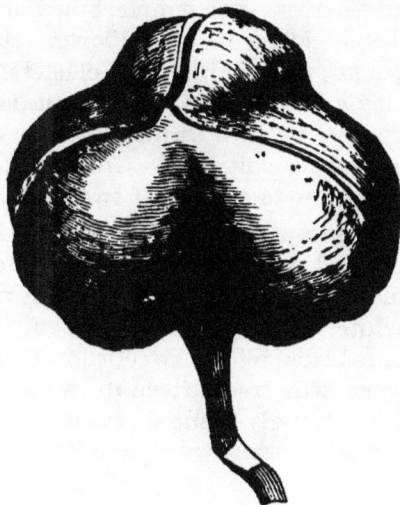

Fig. 49.—SPANISH BUCKEYE.

UNGNADIA, Endl.—*Spanish Buckeye.*

A genus of only a single species, but closely related to the Horse-chestnuts or Buckeyes. It is quite remarkable for having the foliage of a Hickory, and flowers and fruit resembling those of the Buckeye.

Ungnadia speciosa, Endl.—Spanish Buckeye.—Leaves composed

of seven opposite, ovate, long-pointed serrate leaflets. Flowers
rose-color, about a half inch broad, either perfect, or the staminate and pistillate separate on the same plant. Flowers in
sufficient numbers to make the plant quite showy in spring.
Fruit in a leathery capsule, the size and form shown in fig. 49.
The kernel of the nut rather pleasant tasted, but unwholesome,
containing marked emetic properties. A small, handsome tree,
twenty feet high, but more commonly a shrub. In Texas and
Eastern New Mexico. Cultivated in the South as an ornamental tree, also in France, but said to be somewhat tender in the
gardens of Paris. Propagated from seeds, suckers, or by grafting on stocks of the common Western Buckeye.

VIBURNUM, Linn.—*Arrow-Wood, Etc.*

A large genus of evergreen and deciduous shrubs, a few are
small trees, with simple, but commonly toothed, and sometimes
deeply lobed leaves. Flowers showy, mostly white, in compound, terminal, flattish clusters. Fruit a drupe, containing a
single flattish seed. The genus is represented by about a dozen
species in the United States, two of which extend entirely
across the continent. Only two or three of our indigenous species grow to a hight of twenty feet.

Viburnum Lentago, Linn.—Sheep Berry.—Leaves ovate, strongly
pointed, very sharply serrate, smooth, the long margined petioles and midrib, sprinkled with rusty-colored glands. Flowers
white, slightly fragrant. Fruit oval, about a half inch long,
blue-black, with a sweetish, rather mealy edible pulp. A handsome little tree, fifteen to twenty feet high, with hard, yellowish, strongly-scented wood. From Hudson's Bay in British
America, southward to Georgia, in moist soils, also west to
Iowa.

V. prunifolium, Linn.—Black Haw.—Leaves broadly oval, obtuse at both ends, finely and sharply serrate, smooth and shining above. Flowers in large sessile clusters. Fruit ovoid-oblong,
black, edible. A common large shrub or small tree, fifteen to
twenty feet high, in dry, rich woods, from the New England
States, south to Florida, and westward to Texas and Missouri.

V. Opulus, Linn.—Cranberry Tree.—Leaves strongly three-
lobed, broadly wedge-shaped or truncate at base, the lobes
pointed and toothed on the sides, entire in the sinuses. Mar-
ginal flowers of the cluster destitute of stamens and pistils, but
many times larger than the other, forming a kind of ray, which

is quite showy. Fruit ovoid, bright red, pulp very acid, but is sometimes used as a substitute for cranberries. This is the parent of the well known Guelder Rose or Snowball of gardens, in which the flowers are all sterile. A handsome large shrub, sometimes twenty feet high in swamps in the Northern States, and westward extending to the Pacific Coast in Oregon and northward. The remaining indigenous species are either small or large shrubs, seldom over ten feet high, but interesting ornamental plants.

XANTHOXYLUM, Linn.—*Prickly Ash.*

A large genus, the species mostly tropical or sub-tropical evergreen trees or shrubs, with minute monœcious or diœcious flowers, with unequally pinnate leaves, and branches armed with prickles or strong spines. Bark, leaves and fruit usually pungent and aromatic. The fruit of one or two Asiatic species is used in China and Japan as an antidote for nearly all kinds of poisons, and one as a substitute for pepper. We have four indigenous species.

Xanthoxylum Caribæum, Lam.—Satin Wood.—Branches and leaf-stalks unarmed ; leaflets five to seven, ovate-lanceolate, on the fertile plant, and elliptical, obtuse, or emarginate on the sterile. Seed solitary, obovate, black and shining. Said to have been discovered at Key West, Florida, by Dr. Blodget, and to be a large and common tree in that region.

X. Clava-Herculis, Linn.—Tooth-ache Tree, Prickly Ash.— Branches and leaf-stalks armed with long prickles. Leaves alternate, seven to nine foliolate ; leaflets ovate-lanceolate, crenate, shining above ; panicles terminal. Fruit rather downy, containing black seed. A small tree, about twenty feet high in Southern Virginia to Florida, but in the West Indies it grows forty or more feet high. Wood yellow, close-grained, and according to Sloane, has the aromatic odor of Sandal-wood. The leaves, bark and fruit have a pungent aromatic taste.

X. Pterota, H., B., K.—Bastard Iron Wood.—Branches very crooked, armed with short, curved spines, and the leaf-stalks winged and jointed. Leaflets seven to nine, only one half to three-fourths of an inch long ; obovate and crenate above the middle. Flowers in axillary clusters. Fruit about the size of a grain of black pepper, containing one smooth dark-brown seed. A small shrub or tree, with very hard, yellow wood. Southern Florida, west to Texas, also south to Brazil.

X. Americannm, Mill.—Northern Prickly Ash.—Leaves composed of four or five pairs of leaflets, and an odd one ; ovate-oblong, downy when young. Flowers minute, yellowish-green, appearing with the leaves. Fruit small, in clusters, red, and ripe in autumn, very pungent tasted, and often used as a medicine, sometimes for tooth-ache, hence one of the common names of this shrub. A large shrub, ten or more feet high, with prickly branches and smooth grayish bark. In rocky woods, often along roadsides. Middle and Northern States.

XIMENIA, Plumier.—*Hog Plum.*

A genus of a few species of small evergreen trees, mostly tropical, with thorny branches, producing handsome plum-like edible fruit. We have only one species.

Ximenia Americana, Linn.—Hog-Plum, Mountain Plum. — Leaves two inches long, oblong-obtuse, short petioled ; peduncles two to four flowered. Flowers small, yellow. Fruit yellowish, round, as large as a plum, edible. Nut round and white. A small tree with yellow wood. Key West, Florida, and through the West Indies.

ZISYPHUS, Juss.—*Jujube.*

A genus of some fifty species, mainly in Egypt and Southern Asia. The *Z. jujube* is widely distributed throughout Southern Europe and Northern Africa, and its dried fruit well known in commerce. The species in general are spiny shrubs or small trees belonging to the *Rhamnaceæ* or Buckthorn Family, and often bearing edible fruit. The genus is represented with us by two species, neither of any especial value. One in Southern California, the *Z. Parryi*, Torrey, growing about fifteen feet high, and another, the *Z. obtusifolius*, Gray, in Western Texas and New Mexico, is sometimes a small tree, twenty feet high, but more frequently a shrub.

CHAPTER XVI.

EVERGREEN TREES.

TAXACEÆ.—*Yew, Torreya, Etc.*

An order of several genera, principally evergreen trees or shrubs, closely allied and usually included in the Coniferæ as a sub-family, but as their fruit resembles the drupaceous, rather than the coniferous, some of our more modern botanists have very properly placed them in a separate group or order preceding the true cone-bearing genera. They are but slightly resinous. Flowers diœcious, the sterile ones in globose catkins, the fertile solitary, axillary, and the fruit drupe-like, with a pulp surrounding, but not always quite enclosing the bony nut-like seed. This order is represented in the United States by only two genera, and four, or at most five species.

TAXUS, Tour.—*The Yew.*

Small trees or shrubs with widely spreading branches and linear, rather flat rigid leaves. Fertile flower, scaly bracted, consisting of a single ovule or cup-like disk, which becomes large and berry-like, surrounding the nut-like seed. There are in all seven recognized species, three belong to the United States, one in Mexico, and the others to the cooler regions of Europe and Asia.

Taxus baccata, Linn.—Var. Canadensis, Gray.—American Yew. —The American or Canada Yew was by the older botanists considered a distinct species from the English Yew, *T. baccata*, but Dr. Gray and others of our times give it no higher rank than a well defined variety. It is in every way quite similar to the English Yew, except it is merely a low, straggling shrub, only three or four feet high. It is common in our Northeastern States, and occasionally along the mountains to Virginia. Leaves about an inch long, linear, numerous, mostly arranged on two rows, but sometimes scattered thickly around the terminal shoots. The fruit consists of a globular, red fleshy disk surrounding, but not quite covering at the top the nut-like seed. The species or English Yew grows to a large-sized tree, and lives to a great age, and is historically one of the most noted trees in the world. There are many varieties, most of which succeed in this country if planted in a half shady position or protected from the scorching rays of the sun. All are readily propagated from either green or ripe wood.

T. brevifolia, Nutt.—Western Yew.—Leaves nearly an inch long, sharp-pointed, the margin somewhat revolute, bright green above, pale beneath, narrowed at the base into a short slender petiole. Fruit amber-red, much flattened. A tree twenty to sixty feet high, with long, slender, somewhat drooping branches. Wood reddish, hard and tough, very elastic, and like that of all of the Yew Family, valuable for many purposes. This species is found in Central California, northward to British Columbia.

T. Floridana, Nutt.—Florida Yew.—Leaves quite narrow and sharp-pointed, with revolute margins, closely resembling those of the last. A small tree on the banks of the Apalachicola River in Florida, and may prove to be only a local variety of the last.

The Mexican Yew, *T. globosa*, is a tender species, and will not thrive in the open air in the mild climate of England, and is of no especial interest to any one except the botanist. The Japan species are more hardy, especially a dwarf one known as *T. adpressa*, which has very small, oval leaves, short-pointed and pale pink fruit. Another, the *T. cuspidata*, of Siebold, is a much taller tree, with larger, rather thick rigid and exceedingly sharp-pointed leaves.

TORREYA, Arnott.—*Fetid Yew.*

A genus of evergreen trees, including four species each, restricted to a locality of limited extent. Leaves larger and longer than those of the common yews, and arranged in single rows. Flowers similar to those of the *Taxus*, and seed enclosed in a fibrous fleshy envelope of a greenish-brown color. This genus was named in honor of the late Prof. John Torrey, of New York.

Torreya Californica, Torr.—California Nutmeg.—Leaves one to three inches long, and about an eighth of an inch broad, nearly flat, sharp-pointed, the petioles somewhat twisted, bringing the blades into two ranks, bright green above, lighter colored beneath. The fruit obovate to oblong-ovate, one inch to an inch and a half long, the fleshy envelope thin and somewhat resinous. Wood light-colored, close-grained, compact, and very fragrant. A large tree of fifty to seventy-five feet high, with

stem one to three feet in diameter.   California, from Mendocino
County to Mariposa County.

T. taxifolia, Arn.—Stinking Cedar.—Leaves about an inch and
a half long, very sharp-pointed, rigid, almost sessile, pale shining green.   Branches horizontal spreading, with somewhat two
rounded branchlets.   Fruit about the size and shape of a nutmeg, with a smooth bark or shell.   A small branchlet is represented in figure 50, about two-thirds of the natural size.   A
small tree, twenty to forty feet high, with odoriferous and very

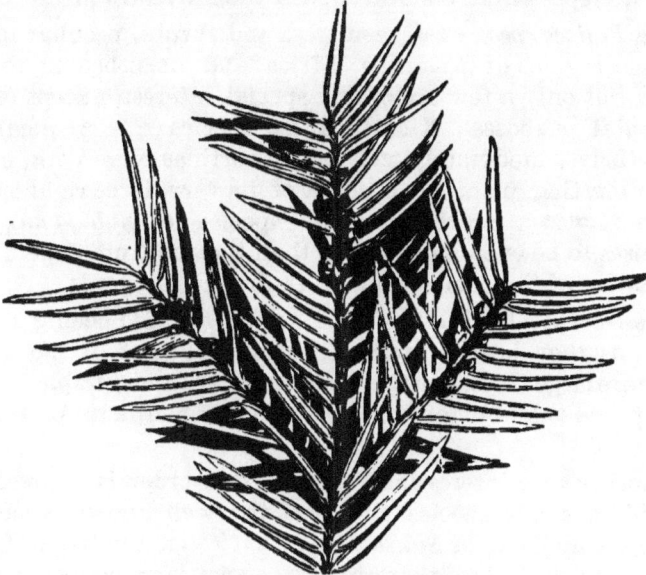

Fig. 50.—FLORIDA TORREYA.

durable wood.   In middle Florida.   This species has proved
quite hardy in favorable soils and locations as far north as the
City of New York, and in a few instances farther north, but
cannot be recommended for general cultivation except in the
South.

FOREIGN SPECIES.

T. nucifera, Zuccarini.—Nut-bearing Torreya.—Leaves as in our
native species, but of a dark glossy-green color.   Branches
numerous, with scaly bark.   Fruit egg-shaped, and about an
inch long.   A tree from forty to sixty feet high, native of the
West Coast of Nippon, Japan.   Not thoroughly tested in this
country, its hardiness is somewhat doubtful in our Northern

States, but will probably succeed south of the latitude of Washington.

T. grandis, Fortune.—Tall Torreya.—Leaves slightly shorter than the last, slightly convex above. Fruit plum-shaped, and about three-quarters of an inch long. This species is scarcely distinguishable from the last. Native of China and the Himalayas. A tree forty to fifty feet high.

The Chinese Yews, or *Cephalotaxus*, of which there are two or three species, would naturally fall into this group, but we have no representatives nearer than the *Torreya's* in our flora.

The *Podocarpeæ*, evergreen trees and shrubs, peculiar to the warmer regions of Australia, Africa, and Asia, belong to this order, but only a few are of any special interest, except for ornamental purposes. The Japanese species (*P. Japonica*) Siebold, thrives moderately well as far north as New York, and is an interesting plant on account of its very dark, rigid leaves. There is also a South American species (*P. nubigœna*), that promises to be even more hardy than the one from Japan, as it is a native of the cool regions of Chili and Patagonia.

The New Zealand Pines, or *Dacridiums*, also belong to this order, as they bear drupaceous fruit, like that of the Yews. Some are large trees, and the wood very hard and durable. They may prove valuable for cultivation in our Southern States and westward.

Another very interesting genus of this order is represented by only a single species, and that the well known Ginkgo or Maiden Hair Tree, or *Salisbury adiantifolia*, a native of China and Japan, and a tree that grows to a very large size, or in some situations a hundred feet high, with stem five to ten feet in diameter. Its leaves are deciduous, fan-shaped, very broad, and cut or notched at the apex. Its fruit is a globular ovate, and an inch in diameter. It is a well known hardy tree, introduced into this country a century ago, or in 1784, by Alexander Hamilton, who planted specimens near Philadelphia, which are said to be still alive and growing. There are several varieties, but none that are really more beautiful than the species.

CHAPTER XVII.

CONIFERÆ, OR CONE-BEARING TREES.

Many volumes have been written, avowedly for the purpose of giving a correct classification of the cone-bearing trees of the world, but the authors of no two of them agree, except as to some of the most simple characteristics of the different genera and species, and the result is, a confusion that may well astound the novice who desires to find an authority at once unimpeachable, and so thoroughly trustworthy, that it may in all cases be quoted without fear of being led into an error. Even in such a simple matter as names of the different species of conifers, authors disagree, and often so widely that no one but a student, or one well versed in the literature of the subject, can possibly reach a satisfactory conclusion as to the identity of any but the oldest and most familiar.

It is true that such European botanists as Tournefert, Lambert, Linnæus, Endlicher, Loudon, Lawson, and the more modern writers like Gordon, Masters, and Veitch, have aided, and in fact have done some good work in elaborating the various genera of which this great Natural Order of plants is composed, but there is yet much material left in an unsatisfactory condition, owing probably in part to the innate difficulties surrounding the subject, and partly to the lack of the scientific knowledge necessary to trace the affinities and relationship of the different species and genera. But I am inclined to believe that much of the confusion that exists in regard to the classification and the names of the different species of conifers, is the result of prejudice and personal opinion, with a desire on the part of each author to set up a standard of his own, which, to be satisfactory to himself, must differ more or less from that established by rival authors. I may be wrong in this matter, but I cannot well attribute the idiopathies of several of the most noted European authors to any other cause. We certainly cannot accuse them of ignorance, or of not being familiar with the writings of others on the same subject, for their works show quite the contrary. Still, when we find men ignoring science, in order to laud a hero as Veitch, Gordon, and nearly all English authors do, in giving the generic name of the Mammoth Tree of California as *Wellingtonia*, instead of the correct one of

Sequoia, we cannot but distrust them in other matters. But such vagaries of authors are not confined to those of our times, for even the revered Linnæus reversed the generic names of the Firs and Spruces, ignoring the classification of those who had lived long before his time. The continental botanists and nurserymen, however, have in most instances retained the older classifications, placing the Spruces under the generic name of *Picea*, and the Firs under *Abies*, while the English and most of our American authors have followed Linnæus, although there can be no question as to its inaccuracy.

The North American Coniferæ have been carefully elaborated in the works of Drs. Gray, Chapman, Engelmann, and other botanists, but our most comprehensive and best special treatise on the coniferæ, is " The Book of Evergreens," by Josiah Hoopes. This is a work that I can confidentially recommend to those who may desire a more scientific description and classification of either the indigenous or foreign species than will be given in the following pages. Owing to the confusion referred to in regard to the classification of our coniferæ, I may in some instances depart from the alphabetical arrangement of the preceding pages, and place the different genera in the order of the relationship instead.

JUNIPERUS, Linn.—*Juniper*.

An immense genus of evergreen trees and shrubs, and the species widely distributed, and in almost every degree of latitude, although principally in the Northern Hemisphere. The wood of all the species is fine-grained, hard and durable, the heart wood usually reddish and fragrant. Flowers diœcious or sometimes monœcious, the small, solitary catkins, axillary or terminal, upon short lateral twigs. Fruit a scaly bracted drupe, and in some species resembling a berry, more than a true cone, usually emitting a strong resinous odor, and containing one to three hard-shelled seeds. Leaves small, scale-like, persistent and rigid. All readily propagated by seeds or cuttings of the small branchlets, also by layers and grafting.

Juniperus Californica, Carr.—California Juniper.—Leaves in clusters of three, short, thick, and mostly acute. Fruit oblong-ovate, of six or rarely four scales, usually one-seeded, and of a reddish color when ripe. A small shrub, or sometimes a tree, twenty to thirty feet high, with rather stout branches. California, in the Coast Ranges, from the Sacramento River southward to San Diego.

Var. *Utahensis*, Engelm., has more slender branchlets. Fruit round and smaller. It inhabits the Sierra Nevada, Southern Utah, and Arizona.

J. communis, Linn.—Common Juniper.—Leaves rather long, linear, awl-shaped, prickly-pointed, upper surface white, glaucous, under one bright green. Fruit small, round, dark purple, covered with a light bloom. A low, straggling shrub or small tree, seldom more than ten or twelve feet high. This species may well be called common, as it is a native of Asia, Europe, and extends entirely across North America. The berries of this species are employed in giving the peculiar flavor to gin, and an oil extracted from them is also used in medicine. There is an immense number of varieties of this species in cultivation, known under such names as Irish Juniper, Swedish Juniper, Spanish Juniper, Large-fruited Juniper, Weeping, Creeping or Prostrate, and many others more or less common in nurseries and ornamental grounds.

J. occidentalis, Hook.—Western Juniper.—A species very much resembling the California Juniper, but the fruit is smaller, blue-black, and the fleshy envelope resinous. A large tree in Oregon, but becoming a mere shrub further south in California. There are several natural varieties. Var. *monosperma*, Engelm., is a small shrub in Texas, west to Arizona, and northward to Colorado. Var. *conjungens*, Engelm., is said to be quite abundant in Western Texas and New Mexico, in fact the two varieties as well as the species appear to be only climatic forms of the common Juniper. The trees are usually crooked and distorted, but the timber is hard and makes excellent fuel.

J. pachyphlœa,—Torr.—This is another of those peculiar western forms of the Juniper, more or less common in New Mexico and Arizona. An exceedingly slow-growing tree, and Dr. Engelmann says that some trees, two hundred years old, have a diameter of only four to six inches, but an occasional specimen is found with a diameter of two to three feet, but these are usually found in rich, rather moist soils, and in sheltered positions.

J. Virginiana, L.—Red Cedar.—Leaves very small, scale-like on the older branches, but larger on the young twigs or branchlets; very numerous, closely imbricated, and of a dark green color. Branches usually horizontal, but in some soils upright, covered with a thin, scaly bark. Fruit small, dark-purple, covered with a whitish bloom. A very common and well known tree, the heart wood of reddish color and very durable. It is largely employed for cabinet work, pencils, fence posts, etc. A very widely distributed species, extending from New Brunswick to Washington Territory, and southward, in the East to Florida, but is said not to have been found in California, and is rare in the Rocky Mountains. An exceedingly variable species, sometimes a tree sixty to eighty feet high, but the most usual size in the Eastern States is between thirty and forty. I consider this the most valuable of all the Junipers, adapted to the climate of the United States, and should take precedence of others for planting in forests or for other useful purposes.

There is a large number of foreign species that thrive in this country, and especially those inhabiting China and Japan, and while they are of interest to the botanist, and are desirable for ornamental plantations, they possess no valuable economic properties not common to our indigenous species.

CUPRESSUS, Tour.—*Cypress.*

A genus of evergreen trees closely allied to the Junipers, but with monœcious flowers, with the aments or catkins terminal, and of a few pairs of opposite scales. The fertile catkins erect on short lateral branchlets, of six to ten thick scales, becoming a roundish woody cone. Seeds acutely angled. Leaves small, scale-like adnate, and appressed, opposite and imbricated.

Cupressus Govenlana, Gordon.—California Cypress.—Leaves bright green, quite small, thick, and without lateral depressions. Cones small, round, a little less than an inch in diameter, and composed of from six to eight scales. A shrub or small bushy tree, six to ten feet high or sometimes more. In the Coast Ranges of California. Not hardy in our Northern States.

C. Macnabiana, Murr.—McNab's Cypress.—Leaves very small, deep green, somewhat glaucous, conspicuously pitted on the back. Mature cones small, round, a little more than a half inch in diameter. A shrub six to ten feet high, with numerous slender branchlets. About clear lakes, and on Mount Shasta, California. Hardy in England, and may thrive in protected situations in our Middle States.

C. macrocarpa, Hartwig.—Monterey Cypress.—Leaves bright green, acute, obscurely pitted on the back, often with a longitudinal furrow on each side. Scales of young cones with foliaceous tips, mature cones clustered on short, stout peduncles, one to one and a half inch long, and nearly an inch in diameter, with five or six pairs of scales, and about twenty seeds to each. A large tree, forty to seventy feet high, with rough bark and widely spreading branches. On granite rocks near the sea, in California, near Monterey, and southward. Hardy in England, and thrives in our Southern States, but tender in the Northern. This species was early introduced into Europe, and from which several varieties have been produced.

CHAMÆCYPARIS, Spach.—*Cypress.*

Trees with the characteristics of the *Cupressus*, but flattened two ranked branchlets, and the small globose cones maturing the first year. The seeds are also less numerous. In very few botanical works are these trees separated from the Cypress, and the reader can take his choice in the name of the genus, and still have excellent authorities for establishing the correctness of either.

Chamæcyparis Lawsoniana, Parlat.—Lawson Cypress.—Leaves small, deep green, with a whitish margin when young, mostly with a gland on the back. Cones small, about a third of an inch in diameter, of eight or ten scales, with the flattened summit terminated by a narrow transverse ridge. Seeds two to four in each scale, and wing-margined. A magnificent and most graceful tree in Northern California and Oregon, in the Coast Ranges, growing one hundred to one hundred and fifty feet high. Wood excellent, white, close-grained, compact and fragrant, and is known by the local name of Oregon Cedar, White Cedar, etc. Its success in our Northern Atlantic States has been rather unsatisfactory, for in some soils and situations it thrives and grows rapidly, while in others near by, it fails, burning in summer and killing back in winter. It succeeds best in a rather moist soil, and very poorly in a dry one.

C. Nutkaensis, Lam.—Nootka-Sound Cypress.—Leaves only one eighth of an inch long, sharp-pointed, over-lapping and appressed, of a very dark, rich green color, very slightly glaucous, without tubercles. Cones small, globular, solitary, with a fine whitish bloom. Scales four in number, shield-shaped, rough, and terminating in the center with a thick, obtuse, straight point. Seeds about three to each scale. Branches spreading or incurved at the ends. A tree sometimes a hundred feet high in Sitka, and southward to the Cascade Mountains on our northwest coast. Hardy in our Northern States, but appears to suffer more from heat and drouths in summer than cold in winter.

11

C. thuyoldes, Linn.—White Cedar.—Leaves very small, ovate, regularly imbricated in four rows, and of a light glaucous-green color. Branches spreading and drooping. Cones very small and clustered. Seeds few, very small, and nearly round. A large tree, forty to eighty feet high, and stem two to three feet in diameter, usually very straight. Wood reddish, light, soft, but fine-grained, and very durable. Used for a great variety of purposes, and always in demand. This tree is always found in

Fig. 51.—TWISTED BRANCHED CYPRESS.

cold, wet lands or swamps, and widely distributed from New England to Florida, and westward to Wisconsin. This species was made the type of the new genus as given above by Spach, but I certainly agree with Mr. Hoopes when he says in regard to this matter, that our "American botanists, however, who have known it from childhood, and whose facilities for close investigation are amply sufficient, refuse to accept the innovation, and consequently retain it in *Cupressus*."

FOREIGN SPECIES AND VARIETIES.

Of these there are quite a large number, but very few if any of them are hardy in our Northern States, but all can be grown in the Southern, as well as in the milder regions of California. One of the most interesting species is the Weeping or Funereal Cypress of Northern China, and described in Robert Fortune's work on the tea countries of China. He says that it grows to a hight of sixty feet, with weeping branches, resembling in this character the common Weeping Willow. Another curious and interesting species, the *C. torulosa*, Don., comes from India, where it grows to a hight of a hundred and fifty feet, with twisted branchlets, somewhat like ringlets. The cones are quite large, and of the size and form shown in fig. 51. This species is held in religious veneration by the natives, and the twigs and fruit are considered a valuable medicine. There are many other species and varieties described by botanists, but are of no especial interest to the practical forester.

LIBOCEDRUS, Endlicher.—*California White Cedar.*

A small genus of only four species, two in South America, one in New Zealand, and one on our Western Coast. It is closely related to the common Arbor Vitæ (*Thuya*). Cones not reflexed, solitary, terminal, and composed of four to six woody coriaceous, concave scales, terminating in a small incurved spine. Seed unequally winged, usually two under each scale. Leaves imbricated in four rows.

Libocedrus decurrens, Torr.—White Cedar of California,—Leaves very bright green, awl-shaped, sharply acute. Cones three-quarters of an inch to an inch long, scaly-bracted at base, oblong, and the lower scales very short. Branches spreading and incurved at the extremities. A very large tree, one hundred to one hundred and fifty feet high, by four to seven in diameter. In general appearance, this tree resembles an Arbor-Vitæ, in fact has been placed among the *Thuya's* in many of our modern botanical works. The wood is soft and light-colored, not durable when exposed to the weather. In the Coast Ranges of Oregon, and southward to San Diego, California. The cultivation of this species in the Atlantic States has not been very satisfactory, but occasional specimens has lived and made a moderate growth without protection, but I cannot recommend it for planting out any where north of Washington.

The foreign species are even more tender than the native one, but are worth cultivating in the South. The *L. Chilensis*, or Chilian Arbor-Vitæ, is a handsome tree from the Andes of Chili, where it grows to a hight of sixty to eighty feet. *L. Doniana*, the New Zealand Arbor-Vitæ resembles our common native species of the Eastern States, but is tender even in the milder climate of England.

L. tetragona, the Alerze of the Chilian's, is a native of Chili and Patagonia, and is the most valuable timber tree of the country, and although introduced into England in 1849, by James Veitch and Sons, they remark in their " Manual of Coniferæ," 1881, that this species " has up to the present time failed in England, and has now become quite scarce."

THUYA, Tournefort.—*Arbor-Vitæ*.

A genus of evergreen trees and shrubs that may be appropriately termed the Shuttlecock of botanists, at least among those of modern times. Even the spelling of the name has been twisted and changed in almost every conceivable way possible, without wholly destroying the word. Drs. Gray and Chapman, also Hoopes, Paxton, Gordon, and several other equally as good botanical authorities, give it as *Thuja*, while Veitch and Sons, in their Manual of Coniferæ, spell it *Thuia*. Masters in his Monograph on the " Conifers of Japan," Watson in Botany of California, and various other authors, spell it as above or *Thuya*. Linnæus in his Systema Naturæ, 1767, and other botanists of his day, and before it, spell the word with a j instead of a y, and while I am satisfied that the weight of authority would certainly establish the j as being the correct orthography, still I prefer using the y, because it accords with the proper pronunciation. But if our botanical authorities are so much at variance in the name of the genus, we must expect a still wider disagreement in regard to the classification or arrangement of the species and varieties belonging to it, or in closely allied genera.

Masters places all the true Arbor-Vitæ's or *Thuyas*, the *Biotas* and *Retinisporas*, under this one generic name of *Thuya*, while the more common arrangement is to divide these into three genera or groups. I am inclined to think the latter is the most convenient one, and that there are good and well defined characteristics that will enable almost any careful observer to separate the species, even if he has no great amount of scientific intelligence to aid him in the work. The American species belong to the first named genera or group, and have monoecious flowers on different branches, the sterile catkins elliptical ovoid, and the fertile ones ovoid and solitary. Cones small, ovoid, with four to six rather thin scales adhering at the base, and covering two flattish seeds, winged all round the margins. Leaves small and scale-like, in four rows on the flat thin branchlets. Only two species in this country.

Thuya gigantea, Nutt.—Giant Arbor-Vitæ.—Leaves acuminate, incurved ovate, somewhat quadrately and closely imbricated, and obscurely glandular; of a bright green, sometimes of a glaucous-green color. Branches and branchlets erect, the latter flattened and very graceful in form. Cones more or less clustered, and slightly longer than those of the next species. A very large and graceful tree, sometimes two hundred feet high, with a stem ten to twelve feet in diameter. Wood white, soft, and easily worked, said to be very durable. In the Coast Ranges and Cascade Mountains of Oregon, and in Northern California. Like most of the evergreens from the Northwest Coast, this tree is often injured by the heat of summer in our Atlantic States, and browned, or the shoots entirely killed in winter.

T. occidentalis, L.—White Cedar, Arbor-Vitæ.—Leaves quite small, rhombic, ovate, imbricated in four rows. Branches numerous, slender, upright, or widely spreading. Cones small, oblong-ovoid, with thin dry spreading, pointless scales. Seed with a broad wing all round. A common and well known tree in low, moist soils throughout Eastern North America. Wood light-colored, compact and durable. Usually a small tree, growing to a hight of thirty to fifty feet. A tree largely employed for screens and ornamental hedges, as it thrives in a great variety of soils. There are many varieties in cultivation, some exceedingly dwarf, others tall and quite slender. The so-called Siberian Arbor-Vitæ of nurseries, is only a compact growing variety of this species. There are several golden-leaved and silver-tipped varieties, one of the latter originated in my grounds some ten years since, and is now in the collection of Parsons and Sons, Flushing, N. Y. I gave it the name of "Columbia," as there is another silver-tipped variety known as "Victoria." But these garden varieties are more interesting as ornamental trees than for practical utility.

BIOTA, Don.—*Oriental or Eastern Arbor-Vitæ.*

Flowers similar to those of the *Thuya*, but leaves small, ovate, scale-like, rough and hard to the feel, imbricated in four opposite rows. The cones elliptic, with thick ligneous, or leathery scales placed in opposite pairs, and furnished with a recurved short or long horny point. Seeds two at the base of each scale, large, ovoid, nut-like and without wings. There is a large number of varieties, by some authors considered species, the most familiar are Chinese Arbor-Vitæ (*Biota orientalis*), a tall growing tree, found throughout China and Japan, and of which there are a large number of cultivated varieties. The Tartarian Arbor-Vitæ (*B. Tartarica*), is probably only a variety of the Chinese, although quite distinct in form of growth and in size and shape of the cones. There are also golden-leaved, weeping, dwarf, and other forms described in works devoted exclusively to the coniferæ, like those of Hoopes, Veitch, Masters, Gordon, etc., etc.

RETINISPORA, Siebold.—*Japan Arbor-Vitæ.*

A genus or group more closely allied to the Chinese than American Arbor-Vitæ, having small, round woody cones, with numerous ovate scales. The seeds are resinous, and with membraneous wings that are usually deciduous, when fully mature. The name of the genus derives its origin from the resinous coating of the seed. There is a very large number of varieties in cultivation, probably all descendents from one original species, but in the present state of our knowledge it would be difficult to fix upon the parent stock. M. T. Masters in Monograph already referred to, names *R. pisifera*, Siebold, and *R. obtusa*, Sieb., as the two species from which the almost innumerable varieties have descended. The last species grows to a large size on the Island of Nippon in Japan, forming trees sixty to eighty feet high. All the species and varieties are really beautiful trees and well worth cultivating for ornamental purposes, if for no other. They present a great variety of foliage, both in form and color. In some the leaves and branchlets are exceedingly minute and feather-like, either dark green or of a silver or golden color, while others have flattish branchlets, somewhat after the forms and character of our common Arbor-Vitæ. Seedlings often vary widely from the parent stock, and what are termed "sports," frequently appear among old and well established plants. One of the most unique varieties in cultivation originated in my grounds about eight years ago, and was described by Prof. Geo. Thurber in the *American Agriculturist*, 1881, under the name of Fuller's Japan Arbor-Vitæ. It originated from a sport of the variety known as *R.* var. *aurea plumosa*, a single branch shooting out from the side of a large plant, and instead of retaining the original form, it pushed out horizontally, and unlike the usual light, feathery foliage, characteristic of the variety, the leaves in this were flat and closely pressed to the stems, presenting altogether, in the form at least, the appearance of a Lawson Cypress. This branch was layered and removed, and is now a tree more than twice the hight and size of the parent plant at its side. The leaves have the golden color or the original, but the plant has the graceful habit of the Lawson Cypress. Furthermore I have found it quite difficult to propagate from cuttings, while, as is well known, the parent is almost as readily propagated in this way as a willow. This freak among the *Retinisporas* in my own grounds has rather lessened than increased my confidence in some of the attempts that have been made to elaborate or correctly classify the different species and varieties of this genus.

SEQUOIA, Endl.—*Redwood—Mammoth Tree.*

A genus of only two species, both of which belong to California. Flowers monœcious, terminal, solitary. Staminate flowers small, partly enclosed with scale-like leaves. Fertile aments, oblong-ovate, erect, the cone maturing the second year, woody, oval, the scales divergent at right angles from the axis, thick and wedge-shaped. Seeds flat, oblong-ovate, with a spongy margin.

Sequoia sempervirens, Endl.—Red Wood.—Leaves a half inch to an inch long, bright green, slightly silvery beneath, spreading in two rows. Cones oblong, only about an inch long, solitary and terminal, with numerous thick, rough scales. Seeds three to five under each scale. One of the most valuable trees in California, occupying the Coast Ranges from Oregon to San Luis Obispo, appearing to thrive best where exposed to the fogs from the ocean. A tree growing from two to three hundred feet high, with a very straight cylindrical stem. Wood a rich brownish-red color, light, but strong and durable, and very straight grained, easily worked, and takes a high polish. Only succeeds in our Southern States, scarcely hardy even in Virginia.

S. gigantea, Decaisne.—Big Tree, Great Tree of California.— Leaves pale green, and much smaller than in the last, not in rows or ranks, slightly spreading or closely appressed, ovate or acuminate, or lanceolate, rigid and pungent. Cones ovate-oblong, two to three inches long, of usually twenty-five to thirty scales. Seeds three to seven to each scale, brownish, with spongy wing-like margin. Cone and a small branch shown in fig. 52, each about one half natural size, with seed between full size of nature. This is called the "pride of the California woods," and it occurs only in groves and isolated groups that extend along a line of some two hundred and forty miles. The largest and tallest one yet discovered, is in what is called the Calveras Grove, and is three hundred and twenty-five feet high. The so-called Grizzly Giant of the Mariposa Grove is a little over ninety-three feet in circumference at the ground. Unfortunately this valuable and noble tree is not a success in our Eastern States, and although an occasional specimen will thrive, it does not appear to be adapted to our climate. I raised a large number of seedlings in 1858, which were distributed among my acquaintances, but I very much doubt if there is one now alive. I have also procured specimens many times since, but sooner or later they would die out. A cool, moist soil, and climate, where the winters are not very severe, appears to suit it best. It seems to thrive well in England.

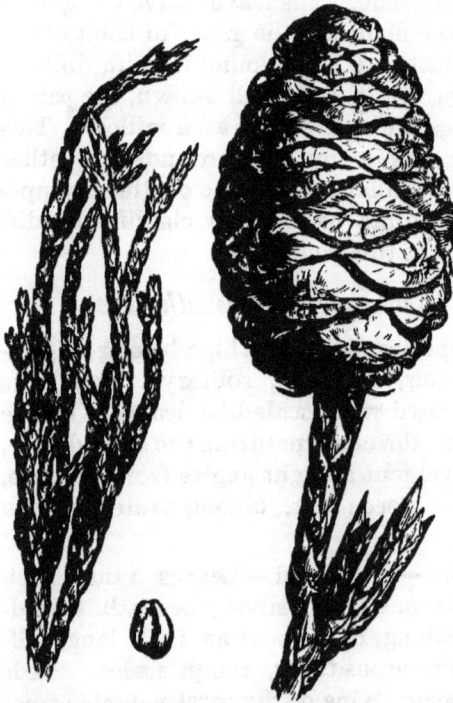

Fig. 52.—CONE, BRANCHLET, AND SEED OF SEQUOIA GIGANTEA.

TAXODIUM, Richard.—*Bald Cypress.*

A genus, as now restricted, containing but one species, and this found in our Southern States, and westward into Mexico. The flowers are monœcious on the same branch. Sterile catkins in a long, spiked panicle, drooping with few stamens, fertile ones with low ovules at the base of each scale. Leaves deciduous, and set in two ranks on the branchlets.

Taxodium distichum, Richard.—Deciduous Cypress, Bald Cypress, etc., etc.—Leaves from one half to three-quarters of an inch long, linear, acute, flat, alternate or opposite, occasionally in whorls. Cones an inch in diameter, round, closed, hard, and rough, with thick woody scales. Seeds small, hard, with narrow wings. While this is a strictly Southern tree, it thrives in all of the Middle, and many of the Northern States. In the alluvial bottom lands of the South, it grows to a hight of one hundred and fifty feet, with a stem ten to twelve feet in diameter. Wood reddish, strong, light, easily split and worked, extensively used for shingles, railway ties and other purposes. It is a rapid growing tree even in our Northern States, and a number of years ago I raised several thousand for stakes, commencing to thin out the young trees when five or six feet high, and I found that it was cheaper to raise stakes on my own grounds than to purchase and haul them ten or twenty miles. This tree deserves more attention from those who are cultivating forest trees than it has ever received. It is a very hardy tree in my grounds, and grows quite rapidly, even in a dry, sandy soil.

There are several ornamental varieties in cultivation, one of a dwarf habit, and another having a very decided pyramidal-shaped top. A Mexican variety differs from the species in having very long persistent leaves, and somewhat larger cones, with the scales armed with a short, stout point.

Before leaving this genus of deciduous conifers, I must refer to another which is so closely allied that several of our botanical authorities have placed the species among the true *Taxodiums*, and classed them under this generic name. I refer to the *Glyptostrobus*, a genus containing at most two species, both inhabiting the colder parts of China and Japan. The *G. heterophyllus*, Endl., is a tree with very small leaves, quite variable in form, scattered all around the branchlets, and of a glaucous-green color. It is only a small tree with ascending branches recurved at the extremities. The other species is known as the " Weeping Deciduous Cypress " (*G. pendulus*, Endl.) It has very slender branchlets, drooping, curved or twisted, and the leaves are long, slender and compressed when young, but spreading at maturity. A hardy and beautiful rapid growing tree, although it probably never reaches a very large size.

ABIES, Tour.—*Fir Tree.*

Evergreen trees and shrubs, with flat, somewhat two-ranked leaves. Flowers monœcious, or male and female on the same plant, but separate ; the male catkins axillary or terminal, the female on very short branchlets. Cones cylindrical, erect, and on the upper side of the branches. The scales of the cones fall from the axis at maturity, not adhering and falling together as in the Pines and Spruces. Seeds with very thin and somewhat persistent wings.

Abies balsamea, Marshall.—Balsam Fir, Balm of Gilead Fir.— Leaves an inch long, or a little less, narrow and slender, spreading, and slightly recurved, dark green above and silvery beneath. Cones three to four inches long, cylindrical. Scales broad, thin, smooth and rounded. Seeds angular, small. A handsome tree when young, but soon loses its lower branches, becoming rather naked and top-heavy. A moderate sized tree, usually growing thirty to forty feet high, but sometimes sixty or seventy. Wood white, soft, and of little value. The liquid resin, known as " Canada Balsam," is obtained from this species. A common tree in cold, damp soils, from Canada southward to Virginia, along the mountains.

A. bracteata, Nutt.—Leafy-bracted Silver Fir.—Leaves two to three inches long, linear, and crowded in two rows, flat, and somewhat rigid, light green above, silvery beneath. Branches in whorls, the lower ones drooping. Cones three or four inches long, and about two in diameter, solitary, with roundish kidney-shaped, rigid, and three-lobed bracts ; the middle one nearly two inches long, slender and recurved, especially those near the base of the cone ; the upper ones nearly straight. A slender, but very tall tree, often reaching a hight of one hundred feet, and sometimes more. Wood like that of all the firs, and of little value. Found in Oregon, and southward in California, in the Santa Lucia Mountains, at an elevation of from three to six thousand feet.

A. concolor, Lindl.—White Fir, Black Balsam.—Leaves two to three inches long, mostly obtuse, but on young trees often long-pointed, two-ranked, pale green, or silvery. Cones oblong, cylindrical, three to five inches long, and about an inch and a half in diameter, pale green or purplish. Scales twice as broad as high, bracts short, enclosed within the scales ; wing of seed oblique and very persistent. Seeds about three-eighths of

an inch long, somewhat triangular and compressed on the
edges. A large tree, seventy-five to a hundred and fifty feet
high, with stem three to four feet in diameter, covered with a
rough, grayish bark. Wood very white, soft, and of inferior
quality. Miners in New Mexico assured me that this tree was
known as the "Black Balsam" in that region, but they could
give no good reason for such a name, as the wood is very white
and the foliage is often of a light silvery color. A common tree
from Northern New Mexico, northward and westward, at elevations of from three to ten thousand feet, and quite abundant
at the highest elevation, in the first named locality. A handsome variety, with leaves incurved upward along the branches,
and known as *A. C.* var. *Parsoniana*, is far more abundant
than the species in the canyons of the northwestern part of Col-
fax County, New Mexico, where I had an opportunity of examining thousands of specimens a few years ago.

A. Fraseri, Pursh.—Fraser's Balsam Fir.—Leaves somewhat
two-ranked, linear, flattened, obtuse or emarginate, whitened
beneath, the lower ones usually recurved, and the upper ones
erect. Cones oblong, one to two inches long ; bracts oblong,
wedged-shaped, short-pointed and reflexed at the summit. A
rather rare little tree, growing thirty to forty feet high in the
mountains of North Carolina and Tennessee, although Pursh,
who first described it, said he found it growing on Broad Mountain in Pennsylvania. A hardy tree, and handsome while
young.

A. grandis, Lindl.—Great Silver Fir.—Leaves short, slender,
flat, one to one and a half inch long. deep-green above and
silvery beneath. Cones three inches long, and about two broad,
cylindrical, obtuse, erect, solitary, of a chestnut-brown color.
Scales very broad, and incurved on the margin. Seeds small,
oblong, with a brittle, thin wing. The largest species of this
genus growing from two to three hundred feet high, with stem
four or five feet in diameter. California to British Columbia,
near the Coast. Wood soft, white, and coarse-grained, but useful for floors, joist, and beams in buildings, but is not durable
when exposed to the weather. A handsome ornamental tree,
but unfortunately many of those that have been distributed
from our nurseries were grafted on some slower-growing stock,
and these failing has led many persons to think that this species would not succeed in our Eastern States.

A. magnifica, Murray.—Red Fir.—Leaves somewhat quadrangu-

252

lar, curved upward, scarcely an inch long, somewhat two-ranked. Cones six to eight inches long, two to three inches in diameter, purplish-brown ; bracts lanceolate-acuminate, and shorter than the very wide scales, which are from one to nearly two inches long, by scarcely an inch high. Seed slender, with broad wings. Readily distinguished from the next by the enclosed bracts. A large tree, two hundred feet and over in hight, with stem six to ten feet in diameter, at elevations of six to ten thousand feet in the higher Sierras. This may be only a local variety of the next species, as it is not abundant, and no forests of it have as yet been found.

A. nobilis, Lindl.—Noble Silver Fir.—Leaves an inch to an inch and a half long, rigid, curved upward, covering the underside of the smaller branches, whitish, and keeled on the upper and under side, rather acute, slightly grooved, and somewhat two-ranked. Cones cylindrical-oblong, six to nine inches long, and two to three in diameter, and almost covered with the reflexed bracts. This tree is also known in Northern California under the name of "Red Fir," and grows to about the same size as the last, but has a much wider range, forming extensive forests at the base of Mount Shasta, California, and northward in the Cascade Mountains to the Columbia River.

A. subalpina.—Engelm.—This is rather a doubtful species, but has been described under various names by different botanists, such as *A. lasiocarpa*, Hook., and *A. amabilis*, Parl., etc., etc., but it is probably only one of the many forms or varieties of *A. concolor*, which is scattered through the sub-alpine regions of the Rocky Mountains of Northern New Mexico, and northward to Oregon.

FOREIGN SPECIES.

Although there are few of these that will ever be planted in this country as forest trees, still there are quite a number that are very desirable for ornamental purposes. The following are among the best known species :

A. Cephalonica, Loudon.—Cephalonian Silver Fir.—Leaves about three-fourths of an inch long, dagger-shaped, sharp and rigid. A beautiful species, from the highest mountains of Cephalonia, and other parts of Greece. A free-grower, and quite hardy in our Northern States.

A. Cilicia, Carriere.—Cilician Silver Fir.—Leaves from one to two inches long, and a tenth of an inch broad, flat, dark-green above, slightly silvery beneath. A very handsome, compact growing tree, from Asia Minor. Moderately hardy, but occasionally the foliage has been browned in winter on my oldest specimen, now fifteen years planted.

A. Nordmanulana, Link.—Nordmann's Fir.—Leaves an inch or a little more in length, flat, incurved, dark, glossy-green above, pale beneath. A handsome large tree, discovered by Prof. Nordmann in the Adshar Mountains, at an elevation of about six thousand feet. It is common in the Crimean Mountains, and those east of the Black Sea. A hardy and highly prized ornamental tree.

A. pectinata, DeCandolle.—European Silver Fir.—Leaves about an inch long, flat, with occasionally an incurved point. A rather unreliable tree for cultivation in this country, and is usually short lived, probably on account of the heat and dryness of our climate.

A. Pichta, Fischer.—Siberian Silver Fir.—Leaves about an inch long, linear and flat, obtuse and incurved at the apex, mostly scattered or crowded, not evenly distributed, very dark green above, paler below. A very compact growing small tree, from the mountains of Siberia. A hardy tree that thrives in almost any kind of soil or situation. There is a variety known as *longifolia*, with larger leaves and more silvery foliage.

A. Pindrow, Spach.—Upright Indian Fir.—Leaves two inches or more in length, two-ranked, occasionally scattered, flat, acute, deep green, slightly silvery on the underside. A very handsome Asiatic species, but does poorly in our hot, dry climate.

A. Pinsapo, Boissier.—Pinsapo Fir.—Leaves less than an inch long, very stiff and sharp-pointed, scattered regularly round the branches. Branches in whorls, and branchlets very numerous. A remarkably handsome tree from the mountains of Spain, and only succeeds in somewhat sheltered situations in our Atlantic States.

A. Webbiana, Lindley.—Webb's Purple-coned Silver Fir.— Leaves an inch and a half to two inches long, mostly two-ranked, linear, flat, and bright glossy-green above, slightly silvery beneath. A large tree from the Himalayas and Nepal. Wood exceedingly fragrant. The leaves often turn brown in summer, owing to the great heat to which they are subjected in our climate.

PSEUDOTSUGA.—Carriere.

A genus of a single species, intermediate between the Firs and Hemlocks. The flowers appear from the axils of last year's leaves. Male flowers in an oblong or subcylindrical stamineal column, surrounded and partly enclosed in bud scales. Female flowers with scales much shorter than the long-pointed bracts. Cones mature the first season, with persistent protruding bracts.

Pseudotsuga Douglassi, Carr.—Douglass Spruce.—Leaves linear, distinctly petioled, mostly blunt or rounded, nearly an inch long on old trees, but a little longer on young, thrifty specimens. Cones two to three inches long, subcylindrical bracts more or less protruding and reflexed. Seeds triangular, convex on the upper side and reddish ; on the lower, flat and white. A gigantic tree, two to three hundred feet high, and eight to fifteen feet in diameter, with thick, brown, deeply fissured bark. Wood reddish or yellow, coarse-grained, heavy and strong, and considered very valuable. Oregon, and throughout the Coast Ranges, into Mexico. One of the largest and most important timber trees in the West. Var. *macrocarpa*, Engelm., has smaller and more acute leaves, and the tree does not grow to as large size as the species. It occurs in the foot hills of the San Bernardino Mountains, California.

TSUGA, Carriere.—*Hemlock Spruce.*

A genus of five species, one in the Atlantic States, two in Western North America, and two in Asia. Male flower a sub-globose cluster of stamens, appearing from the axils of last year's leaves. Female catkins terminal on last year's twigs, with bracts somewhat shorter than the scales. Large trees, with very slender drooping terminal branches.

Tsuga Canadensis, Michx.—Hemlock.—Leaves linear, a half inch long, flat, obtuse, dark green above and whitish beneath. Cones three-quarters of an inch long, oval, composed of a few roundish, oblong, thin scales. Seeds quite small, with thin wings. A large and most graceful tree, with a light spreading spray of delicate foliage. It grows to the hight of nearly a hundred feet, with stem three to six feet in diameter. Wood light-colored, very coarse-grained, but extensively employed for roof boards and sheathing, as it holds a nail well, also for joists and smaller timber used in buildings. It is inferior in quality to that of the Pines and Spruces, still it is so abundant and cheap that it is largely used for the purposes named. The bark

is rich in tannin, and it is in great demand for tanning leather.
A strictly Northern tree, succeeding only in cool climates.
Very abundant in the Northern States and Canadas, and along
the mountains southward to Georgia.   There are quite a number of varieties of the Canada Hemlock in cultivation, some
with broader leaves than the species, and others with smaller
and deeper green, and several of a dwarf habit, and very compact growth, but the most unique of all is Sargent's Weeping
Hemlock, a very graceful tree, with pendulous branches.

T. Mertensiana, Bongard.—California Hemlock.—Leaves about
three-fourths of an inch long, flat, obtuse, crowded, bright
green above, slightly whitish below.   Branches and twigs very
slender, drooping. Cones about an inch long, ovate, with a few
persistent kidney-shaped entire scales.   A large tree, one to two
hundred feet high, with a more rounded conical head than our
Eastern Hemlock, but wood quite similar, although claimed
by some to be of better quality.   Said to be hardy in England,
but the foliage often burns badly in our Eastern States during
the hot weather in summer.   Native of California, and northward to Alaska, in the Coast regions.

T. Pattoniana, Engelm.—Patton's Hemlock.—Leaves mostly
convex or keeled above, somewhat sharp-pointed, about an inch
long.   Cones cylindrical-oblong, two to three inches long,
seeds larger than in the last species, and the wings shorter.   A
very tall, strictly pyramidal tree, one hundred to one hundred
and fifty feet high, and growing at elevations of eight to ten
thousand feet in the Sierra Nevada and northward through the
Cascade Mountains in Oregon.

The Asiatic species of the Hemlock have frequently been introduced, and so long as kept in a conservatory or carefully
protected in winter, they thrive, but do not succeed when
planted out and exposed to our severe climate.

PICEA.—*Spruce.*

A genus of about a dozen species peculiar to the mountainous
regions of America, Europe and Asia.   Male flowers axillary
or sometimes terminal on last year's branchlets. Female catkins at the end of short or long twigs, with scales much larger
than the bracts.   Cones maturing the first year, and pendulous.
Scales and enclosed bracts persistent on the axis, the cones falling off entire after the seed have dropped out.   Leaves usually
keeled above and beneath, disposed somewhat spirally all

around the branches, seldom in rows or ranks as in the Firs. We have five native species.

Picea alba, Michx.—White Spruce.—Leaves needle-shaped, four-angled, one half to an inch long, and distributed all around the branch, those on the underside curving upward; of a light silvery-green color. Cones one to two inches long, oblong-cylindrical, with entire scales. Seeds small, with thin wings, about three-eighths of an inch long. A very beautiful tree, especially while young. A rather small tree, but sometimes fifty feet high. Native of the northern portion of the United States, extending far northward into British America. Wood light-colored, rather tough and flexible, sometimes used for masts and spars for boats and small vessels on our lakes. There are a few handsome cultivated varieties, the best known are the Blue Spruce (var. *cœrulea*), with dark bluish-green leaves, and the " Glory Spruce " (var. *aurea*), with golden-tinted leaves.

P. Engelmanni, Parry.—Engelmann's Spruce.—Leaves nearly an inch long, strongly keeled below, abruptly, but not sharp-pointed. Cones one and a half to two inches long, and about three-fourths of an inch in diameter, ovate-cylindrical, and very much scattered on the tree. Scales rhombic, with upper ends appearing as though broken off.

In general outline this species resembles the next, but grows to a larger size, or from sixty to one hundred feet high, with stem two to three feet in diameter. Wood white, soft, or in very old trees, reddish and rather coarse-grained, resembling that of the Red Spruce of the Eastern States. In Northern New Mexico, Arizona, Colorado, and northward to British Columbia, in the mountains at high elevations, often reaching up to the very border of what is termed "timber line," or between eleven and twelve thousand feet.

P. nigra, Poiret.—Black or Double Spruce.—Leaves very short or about a half inch long, stiff and somewhat quadrangular, very dark green. Cones from an inch to an inch and a half long, ovate, or ovate-oblong, dull reddish brown when mature. Scales very thin, roundish, with an uneven margin. Seeds small, with rigid wings. A large tree, seventy-five feet high, sometimes higher in deep woods. Wood light-colored, but sometimes reddish, light, strong, well known in all of our Northern States under the name of Red or Black Spruce timber, and lumber of various forms. The Red or Black Spruce (as it is equally well known under both names), is not a handsome evergreen under cultivation, as it soon loses its lower branches and becomes rather a ragged and unsightly tree.

P. pungens, Engelm.—Silver Spruce.—Leaves about an inch long, rather broad, rigid, stout, sharply acute, usually incurved, pale green above, and silvery-glaucous below. Cones three to four inches long, cylindrical and pendulous, as in all of the true species, very abundant, with elongated rhombic, truncate scales. Seeds small, with somewhat triangular obovate wings. This species was formally considered as only a variety of *P. Menziesii,* Douglass, but has recently been raised to the position of a species, and the Menzies' Spruce placed as a synonym of the next. A large and beautiful tree in Colorado, Wyoming and Idaho, but no where in great abundance. Succeeds admirably in the more Northern of our Atlantic States.

P. sitchensis, Bongard.—Sitcha Spruce.—Leaves a half inch or more in length, flat, with a sharp point, whitish on the upper surface when young. Cones cylindrical, oval, one and a half to two and a half inches long, and about one inch in diameter, pale yellowish. Bracts rigid, lanceolate, and about one half the length of the oblong-rounded scales. A large tree, one hundred and fifty to two hundred feet high, with stem five to nine feet in diameter. Wood said to be superior to any other species of the Spruce. Peculiar to the Northern Pacific Coast, mainly in wet, sandy soils near streams in Mendocino County, California, northward to Alaska.

FOREIGN SPECIES AND VARIETIES.

Among these the best known is the Norway Spruce (*P. excelsa*), which has long been a favorite ornamental tree in this country, and probably more extensively planted than any other conifer. It is really a handsome tree, and being a native of Northern Europe and Asia, it is quite hardy in all of our Northern States, except, perhaps, on the western praries, where the winds are more injurious than low temperatures. There are an immense number of varieties in cultivation, in fact, more than I can spare room to name, and for this reason must refer the reader to the catalogues of nurserymen or special works on the coniferæ, for names and descriptions.

There are, however, several other foreign species and varieties not so well known as the Norway Spruce, but equally worthy of cultivation, and among them I will name the

P. Orientalis.—Oriental Spruce.—A beautiful tree, with very short, dark green leaves, about half an inch long, which entirely surround the branches. A regular conical growing tree, but not a rapid grower.

P. polita.—Tiger's Tail Spruce.—A native of the mountains of Japan, is very distinct, with strong, rigid, sharp-pointed leaves, somewhat sickle-shape, on sturdy, strong branchlets, with very prominent buds, as shown in figure 53. Cones four or five inches long, of the shape shown in figure 54. It is a rather slow-growing

Fig. 53.—BRANCHLET OF TIGER'S TAIL SPRUCE.

Fig. 54.—CONE OF TIGER'S TAIL SPRUCE.

species, but of a very sturdy habit. My oldest specimen, ten years planted, is only about eight feet high. This spruce would make an excellent hedge plant, owing to its sharp-pointed rigid leaves.

P. firma is another Japan species from the mountains of Japan, that promises to be a valuable addition to our list of Eastern conifers, but none of these and several other species and varieties that have been introduced from abroad, are abundant enough as yet to be admitted into a list of available forest trees.

LARIX, Tournefort.—*Larch.*

A genus of deciduous cone-bearing trees, closely allied to the Firs (*Abies*), but distinguished by smaller cones, with persistent scales and bracts; usually erect, on slender, rather drooping branches. Sterile flowers, nearly as in the Pine, but pollen grains solitary and round. Fertile catkins lateral and scattering, bright crimson when in bloom. Leaves slender, soft, deciduous, mostly in clusters or bundles at the ends of the short, undeveloped branches. Only about a half dozen species, and these confined to the Northern Hemisphere, but extending entirely around the world, through Asia, Europe, and North America.

Larix Americana, Michx.—American Larch, Tamarack, Hackmatack.—Leaves from one half to three-fourths of an inch long, slender and thread-like, light bluish green. Cones about an inch long, ovoid, scales few, slightly reflexed and rounded. Seed small, with short, thin wings. Branches slender and drooping, and the tree while young has a very graceful habit, but as they grow older the lower branches die, and break off, and the persistent cones adhere to those above, until the trees seem to be loaded down with them, and they are quite conspicuous and not very ornamental, during the winter months. A handsome ornamental tree while young, but soon becomes too tall, slender and naked, as the lower branches soon cease to enlarge or lengthen. A large tree in the cold northern woods and swamps, sometimes reaching a hight of a hundred feet, with a stem two feet in diameter. Always a slender tree, with light colored, strong wood, which is moderately durable, and used in ship building, posts and fencing. The quality of the wood depends somewhat upon the soil or locality where grown, that from British America, Labrador. and Newfoundland, is said to be much superior to that grown within the United States. The Larch is of little value on dry soils, and we have many far more valuable trees for cultivating in moist ones.

L. Lyallii, Parlat.—Lyall's Larch.—A smaller species than the last, found a number of years ago in the Cascade Mountains of Washington Territory, by Dr. Lyall, and described in the "Gardener's Chronicle" by Professor Parlatore. A small tree, growing

only thirty to forty feet high, and remarkable on account of
the cobweb-like wool that clothes the leaf-buds and young
shoots. Its cones are larger and more oblong than those of our
other native Larches. It is found at elevations of six to seven
thousand feet.

L. occidentalis, Nutt.—Western Larch.—Leaves a little less
than an inch long, thick, and quite rigid, terminated with a
sharp point, doubled channeled above and below, somewhat
four-angled, but flat. Cones ovoid, an inch and a quarter long,
reflexed, scales short, ovoid, edges thin. Bracts a half inch
long, fringed, and terminating in a long awn. A large tree,
sixty to eighty feet high in Oregon and Washington Territory,
where it grows up to an elevation of some five thousand feet.
This species will probably thrive in our Atlantic States.

FOREIGN SPECIES AND VARIETIES.

The common European Larch (*L. Europea*), has long been a
favorite forest tree in Europe, not only on account of its valuable timber, but because of its rapid growth under cultivation.
It is found abundantly through Central Europe at high elevations, where it grows to a large size, sometimes a hundred feet
high.   During the past two centuries extensive Larch plantations have been established in Great Britain, especially in Scot-
land, where this tree appears to thrive as well as in its native
mountains.   Its timber is extensively used for naval purposes
on account of its lightness, toughness, and durability. It is
also employed for hop-poles, mill-work, beams, joists in buildings, docks, and various other purposes. The cultivation of
the European Larch in this country has often been attempted
on quite a large scale, and at one time it was thought that it
would prove a valuable tree for planting on the high and dry
prairies of the west, but the climate of those regions does not
appear to be as congenial as that of Great Britain, and, upon the
whole, the Larch plantations in the west have not been as great
a success and was expected, although the tree thrives in almost
any good and moderately moist soil in our Northern States, but
is scarcely adapted to planting on the higher and drier plains
and prairies.   There are several handsome ornamental varieties
in cultivation, and they may be found described in nurserymen's catalogues under such names as the Weeping Larch,
Smooth-leaved, Compact or Pyramidal, etc.

L. Dahurica, Turz.—Dahurian Larch.—A small tree from

Northern Siberia, growing on the bleak mountains of Dahuria, also found in the Ural Mountains and Kamtchatka, to the Pacific Ocean. It is closely allied to the European Larch, and may only be a northern form of the same species.

L. Griffithiana, Hook.—Sikkim Larch.—This was discovered by Dr. Hooker, and as growing in Bhotan, Sikkim, and Nepal, at elevations of six to twelve thousand feet, it is a large, sprawling, irregular growing tree of some fifty or sixty feet high, with rather long leaves, and cones two to two and a half inches long.

L. Leptolepis, Gordon.—Japan Larch.—This is a very handsome species from the mountains of Northern Japan, where it

Fig. 55.—GOLDEN LARCH (*L. Kœmpferi*).

grows to a hight of forty feet. The leaves are an inch to an inch and a half long, slender, and of a pale green color. Cones about an inch and a quarter long, with about sixty scales. Young branches smooth, with ash-colored bark, rather rigid, and spreading branchlets. A very handsome, erect growing tree, and very hardy, at least I have never seen a twig injured by cold in my grounds.

L. Kœmpferi, Gordon. — Golden Larch. — A very distinct species from China, and by some botanical authorities placed in a genus by itself, under the name of *Pseudolarix*, or False Larch. It inhabits Northeastern China, at elevations of about three thousand feet above the sea, where it grows to a hight of a hundred and twenty feet, with a stem three feet in diameter. The leaves grow in bundles, like the common Larch, but one to two inches long, and the cones nearly three inches long, with thick, woody, somewhat divergent scales. The leaves in spring are of a pale pea-green color, becoming darker in summer, and changing to a bright golden color in autumn. A catkin bearing twig is shown in fig. 55, the leaves somewhat reduced in size. This is as yet a rather scarce tree, in both European and American gardens, although it was introduced into England in 1852, and soon after into the United States, but owing to the difficulty of procuring seed, and propagating by other means, the number produced has been quite limited.

PINUS, Tournefort.—*Pine.*

An extensive genus of evergreen trees, containing a larger number of species than any other of the coniferous group. There are in all between sixty and seventy species described in botanical works, eleven of which belong to our Atlantic States, fifteen to the Rocky Mountain regions, and westward to the Pacific, and about the same number to Mexico and the West Indies—the remainder to the Old World, extending from Great Britain to China and Japan. Some of the species thrive in the poorest and lightest soils, which are almost worthless for agricultural purposes, while others grow on rocky cliffs and in bleak and exposed situations or among stone that are merely covered with a thin film of vegetable matter. The genus as a whole may be said to contribute more to the comfort, welfare and prosperity of civilized man than any other order or class of forest trees, while occupying the least valuable portions of the earth's surface.

Flowers monœcious, male catkins exceedingly numerous in spikes or clusters, female catkins solitary, or several together, and scales much longer than the bracts. Fruit a cone, maturing the second year, spreading or reflexed, rarely erect, and composed of woody imbricated scales. Seeds nut-like, situated in an excavation or depression at the base of the scales, mostly winged, but the wings only persistent in a few species. The cones of many of the species remain attached to the branches until they decay and fall to pieces when several years old. Leaves needle-shaped, cylindrical or somewhat triangular, in clusters of two, three, or five, enclosed in a thin sheath at the base. The number of leaves in a sheath not only aid in separating and determining the different species, but the practical forester knows that there is a great difference in the character of the wood of the species belonging to the different groups or divisions. Those with five leaves in a sheath, like our common White Pine, have much finer grained and softer wood than those with two leaves, and so far as my personal observations have extended, this holds good with the Pines of all countries.

Pinus Arizonica, Engelm.—Yellow Pine.—A new species of which little is known, probably Mexican, but collected in Southern Arizona in 1874, by Dr. Rothrock, and described in Wheeler's Reports. Said to be a small tree, growing forty feet high, and yielding the best lumber of that region of country, which is certainly not very high praise, as there are very few valuable lumber trees in Southern Arizona.

P. Australis, Michx.—Long-Leaved Pine, Southern Yellow Pine, Georgia Pine.—Leaves three in a sheath, ten to fifteen inches long, bright green, and somewhat crowded at the ends of the branches. Cones six to ten inches long, cylindrical, with thick scales and very small recurved spines. A large and common tree throughout the Southern States, growing sixty to eighty feet high, with stem three to four feet in diameter. Wood hard, fine grained and durable, extensively employed in ship building, floors, fencing, and inside finishing of buildings. Sometimes containing so much resin as to be of little value, except for burning and making lamp-black. From this species the greater part of the turpentine, tar, pitch, and resin produced in this country is obtained. This tree thrives in the poor, light soils of the South, but is not hardy in the North, although I have known specimens to live for several years in the suburbs of New York.

P. Balfouriana, Jeffrey.—Fox-Tail Pine, Cat-Tail Pine, Hickory Pine.—Leaves in fives, an inch to an inch and a quarter long, rigid, and usually curved or twisted, crowded and appressed to the stem, and remaining on the branches ten or more years. Cones three to four inches long, dark purple or brown when ripe, and usually attached to long, slender branchlets. Scales thick, with short, very brittle prickles. Seeds small, whitish, with wings three-fourths of an inch long. Wood reddish, hard, tough, and close-grained, very durable, and that from slow-growing old trees almost equal to Red Cedar. Quite a variable

species, sometimes a wide-spreading, open-headed tree, with
long, flexible, drooping branches, while other trees near by will
assume a pyramidal form, or even fastigiate, the latter form
more abundant on the dry, rocky sides of canyons in New
Mexico than I ever found it elsewhere. A small tree, seldom
over fifty feet high, with stem three or four feet in diameter,
in California at elevations of five to eight thousand feet, forming extensive forests, also in the high mountains, eastward
through Southern Utah, Colorado, and southward to New
Mexico, growing at elevations of from seven to twelve thousand feet, or up to what is called timber-line. Var. *aristata*,
Engelmann, is described as having more ovate cones with thinner scales and shorter recurved or awn-like prickles. The species, however, is so variable, that a large number of varieties
can be easily found in the region named, and I have often regretted, when examining them in their native habitats, that I
could not transplant some of them to my garden in New Jersey. A few specimens that I sent home at the time of my last
visit to the mountains, two years since, have lived, and were
not in the least injured by the cold of the past two winters, and
I am inclined to think that this very distinct western pine will
succeed in our Eastern States if planted in a light, dry, or well-
drained soil.

P. Banksiana, Lam.—Gray Pine, Scrub Pine.—Leaves in twos,
from a very short sheath, only an inch long, quite rigid, and
evenly distributed, and of a grayish-green color. Cones about
two inches long, ovate-conical, curved or bent to one side,
smooth, of a light gray color, scales almost or quite pointless.
A small, low tree, twenty feet high, or only a low, straggling
shrub. Common far North, and barely reaching our northern
borders in Maine, Michigan, and westward to Dakota.

P. Chihuahuana, Engelm.—Chihuahua Pine.—This is another
Mexican Pine that barely extends across the line into Southern
Arizona, on the mountains. A small tree, growing thirty to
forty feet high, and of little value, except where wood is quite
scarce.

P. contorta, Dougl.—Twisted-Branched Pine.—Leaves in pairs,
an inch to an inch and a half long, strongly and closely serrulate. Cones clustered, oval or cylindrical, two to two and a
half inches long, scales smooth, or furnished with a very delicate prickle. Two cones and a pair of leaves are shown in figure 56, cones somewhat reduced in size. Illustration from

"The Garden," London, accompanying an article on this Pine by Andrew Murray, Esq. Cones often remain closed for a year or two after they are mature. A small tree, rarely more than thirty feet high, with wide spreading and somewhat twisted branches. Wood light-colored, straight-grained, but usually too small to be of much value. A tree found in swampy grounds near the sea coast, from California northward to Alaska. Var. *Murray-ana*, Engelm., is a much taller-growing tree, sometimes reaching a hundred feet high, and stem four to six feet in diameter, with longer leaves and cones, opening at maturity, all of which may be due to a more favorable soil and climate, as it is found in the higher Sierra Nevada, eastward to Utah, Colorado, and Northern New Mexico, but an occasional specimen will be met in these regions, cor-responding in almost every

Fig. 56.—LEAVES AND CONES OF PINUS CONTORTA.

particular with the description of the species as it is found on the Northern Pacific Slope. Both species and variety succeed in our Atlantic States.

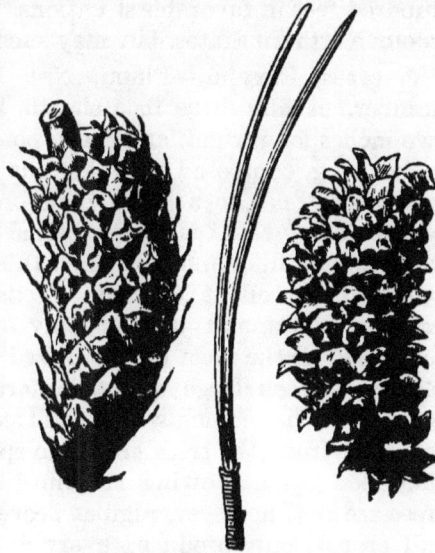

P. Coulteri, Don.—Coulter's Pine, Hooked-Cone Pine.—Leaves in threes, six to eleven inches long, quite large and coarse. Sheath an inch and a half long while young. Male flowers cylindrical and almost or quite two inches long, surrounded by eight or ten bracts, Cones very large, on short stems, long, oval-pointed, ten to fourteen inches long, and four or five in diameter, of a yellowish-brown color, each scale terminated by a long, very strong incurved point, in some instances this horn-like point is two inches long. Seed oval, dark-colored, nearly black, and a half inch or more in length. Nuttall says that this tree was first discovered by Dr. Coulter on the Santa Lucia Mountains, near the Mission of San Antonia, in the thirty-sixth degree of latitude, and within sight of the sea, at an elevation

12

three to four thousand feet above it. Found in California, only in the Coast Ranges, principally in the southern part of the State. I obtained cones of this species, and several others growing in the same region some twenty odd years ago, at a cost of ten dollars each, but have no personal acquaintance with the wood, but it is said to be brittle. The tree reaches a hight of a hundred feet in favorable situations. Too tender for cultivation in our Northern States, but may succeed south of Washington.

P. edulis, Engelm.—Pinon, Nut Pine.—Leaves variable in number, usually three in a sheath, but often only two ; about two inches long, rigid, and sharp-pointed. Cones two to three inches long, composed of numerous small scales at base, and a few larger ones, nearly an inch broad above. Scales blunt, with a yellow reflexed resin-covered tip. Seed a half inch or more in length, cylindrical, shell thin and brittle, kernel white, sweet, and excellent flavored. Seeds two at the base of the uppermost scales, and usually only one in the lowest fruiting row. This is the most highly prized of all the nut pines for its seeds, of which large quantities are gathered by the Indians residing in the regions where the tree abounds. The cones are whipped from the trees and then spread out in the sun, where they soon open, allowing the nut-like seeds to drop out. The trees are not, however, regular bearers, and in some localities a full crop is only produced every five to seven years. A low-growing tree, twenty to thirty feet high, with a stem a foot in diameter. Wood most excellent fuel. In groves, or scattering along the dry banks of canyons, and in stony soils, from Colorado, through New Mexico and Arizona. Hardy in our Northern Atlantic States, but foliage sometimes burns in summer.

P. Elliottii, Engelm.—Elliott's Pine.—A species said to be more or less common near the coast in South Carolina and Florida, growing among and often confounded with the common Old Field Pine (*P. Tœda*). It may prove to be the *P. Tœda*, var. *heterophylla*, of Elliott, described in his Botany of South Carolina, Vol. II, p. 636. Leaves not of a uniform number in a sheath.

P. flexilis, James.—Western White Pine.—Leaves in fives, two to two and a half inches long, somewhat rigid and triangular, sharp-pointed, and densely crowded on the branchlets, of a rich, dark-green color. Cones cylindric tapering, four to six inches long, and two to three in diameter. Scales thick, an inch and a quarter broad, woody, and of a greenish-yellow color when mature. Seeds rather large, irregular obovate, with firm-keeled margins. A handsome tree, resembling the White Pine of the Eastern States, but of a more compact habit, and the foliage darker green. It grows fifty to sixty feet high, with a very straight stem and smooth bark, until the trees become old. Wood white, soft, and easily worked, closely resembling the White Pine of the East. This species inhabits the mountain ranges from Montana to New Mexico, Arizona, and on the Inyo Mountain in California, at high elevations, or from eight to ten thousand feet. Var. *albicaulis*, Engelm., is a smaller tree with more oval cones, and not quite as long, thicker and somewhat pointed scales. An alpine form found in Montana and British Columbia, also in some of the mountains of California. A handsome and hardy tree, worthy of extended cultivation.

P. glabra, Walter.—Spruce Pine.—Leaves in twos, three to four inches long, slender, scattered. Cones about two inches long, solitary, spines nearly obscure ; wings of seed light-colored, long and tapering. Branches and branchlets smooth and light-colored, or whitish. A tree forty to sixty feet high, with soft, white wood. A somewhat rare tree in swampy grounds through South Carolina, Florida, and westward.

P. inops, Ait.—Jersey Pine, Scrub Pine.—Leaves in twos, and from two to three inches long, from a short sheath, scattered, rigid, and flat on the inner surface. Cones light-brown, oblong-ovoid, two to three inches long, often curved to one side. Scales armed with a straight, strong spine. The cones open when mature, allowing the small-winged seeds to fall out. Branches spreading and flexible, covered with a smooth, whitish bark while young, but becoming dark-colored and rough with age. A small tree, fifteen to forty feet high. Wood of little value except for fuel. A widely distributed species on Long Island, Staten Island, New Jersey, and southward to Florida.

P. insignis, Dougl.—Monterey Pine.—Leaves in threes, four to six inches long and very slender, very closely serrate, bright green. Cones on short stems, in clusters, deflexed, three to five inches long, and two to three in diameter ; deep chestnut-brown, persistent, and remaining closed for several years. Scales near the base, very thick and roundish. Seeds grooved and rough, black, about a quarter of an inch long, with wings nearly an inch long, broadest above the middle. A large tree, eighty to one hundred feet high near the coast in California, south of San Francisco. A rapid-growing tree, with a beautiful fresh green foliage, but tender, except in our Southern States.

P. Lambertiana, Dougl.—Lambert's Pine, Sugar Pine.—Leaves in fives, three to four inches long, from short deciduous sheaths, with five or six lines of stomata on each side. Cones twelve to eighteen inches long, and three or four in diameter, gradually tapering to a point on peduncles three inches in length, pendulous when mature, and of a brown color, destitute of resin. Scales loosely imbricated, rounded above, without spine or prickle. Seeds oval, nearly a half inch long, kernel sweet; wing almost twice as long as the seed, of a dark color. A very large tree, one hundred to three hundred feet high, and ten to twenty feet in diameter, with branches in whorls, bark smooth and light-colored, except on the stem and larger branches. Wood white, soft, resembling that of the White Pine, but a little coarser-grained. More or less abundant throughout California and northward to the Columbia River, on both slopes of the Sierra Nevada, and at elevations of from three to eight thousand feet. The exudations from the partly burned trees acquires a sweetish taste, whence the name of "Sugar Pine." A valuable forest tree, and seems to be as hardy in my grounds as the common White Pine, which it very much resembles while young, but when well established, grows far more rapidly, becoming rather tall and naked in appearance, unless the leading shoots are headed back.

P. mitis, Michx.—Yellow Pine, Short-Leaved Pine.—Leaves in twos, three to five inches long, with long sheath, slender, somewhat channelled, and of a dark green color. Cones oval or oblong, about two inches long, usually solitary, with a short, incurved spine on each scale. Seed very small, with a reddish wing. A tree forty to fifty feet high, with stem one to two feet in diameter. Wood yellow, hard, durable, and employed for ship building, spars, masts, plank, etc. In New Jersey, and southward to Florida, also in Missouri and Arkansas.

P. monophylla, Torr. and Frem.—Fremont's Pine, Nut Pine.— Leaves, one or two in a sheath, from one and a half to two and a half inches long, when in pairs, flat on the inner side, single ones round, very rigid, and sharp-pointed. Leaves on terminal branchlets, often bluish, glaucous-green or silvery. Cones two inches long, or a little more, nearly round, of a light brown color, scale thick, recurved, without spines. Seed quite large, wingless, and kernel sweet, edible, used for food by the Indians. A small tree, twenty to thirty feet high, with stem twelve to eighteen inches in diameter, but often only a low, straggling bush. Wood white and soft, resinous, making good fuel. In the Coast Ranges of California, Arizona, Southern Utah, and Nevada. Plants raised from seed, from the higher mountains of Nevada, have proved perfectly hardy in my grounds, neither receiving protection from the sun in summer. Plants of slow growth, but are unique, differing widely from all the other species of pine with which I am familiar.

P. montirola, Dougl.—Mountain Pine.—Leaves in fives, three to four inches long, obtuse, smooth, glaucous-green. Cones cylindrical, slender, four to eight inches long, yellowish-brown, with loosely imbricated, pointed, but spineless scales. Seed small, with large wings. A tree sixty to eighty feet high, and sometimes three feet in diameter. A species closely allied to the White Pine, and resembles it in growth, leaves and wood. California, in the Sierra Nevadas, and northward to Washington Territory, at elevations of from seven to ten thousand feet. Hardy, and thrives in light, sandy soils, better than in those that are moist and heavy.

P. murirata, Don.—Bishop's Pine.—Leaves in pairs, four to six inches long, quite broad, rigid, and strongly serrulate, and of a bright-green color. Cones sessile, about three inches long, ovate, in clusters, crowded with thick, wedge-shaped scales, with stout, short prickles. The cones are very persistent, remaining on the trees for many years, and the scales remaining closed for a long time. I have cones of this species in my cabinet, gathered twenty years ago, and although kept in a warm room, only a few of them have opened sufficient to show the seed. A medium sized or large tree, varying in hight in different regions, from twenty-five to over a hundred feet high, with reddish-brown, roughish bark. In California, only near the coast, where it is exposed to the wind and fogs of the ocean, and principally in swamps and wet soils.

P. Parryana, Engelm.—Parry's Pine.—Leaves three to five in a sheath, mostly four, an inch to an inch and a half long. Cones sub-globose, an inch and a half to two inches long, thick, with strongly elevated knobs. Seed oval, about a half inch long, with a thin, light brown mottled shell. A small tree, twenty to thirty feet high, collected only by Dr. C. C. Parry, forty miles southeast of San Diego, across the border in Mexico, and at an altitude of two or three thousand feet. This species is unknown to me, and the above description is taken from Botany of California, Vol. II, p. 124.

P. ponderosa, Dougl.—Yellow Pine, Heavy Wooded Pine.—Leaves in threes, five to nine inches long, broad, coarse, twisted, flexible, and of a deep or grayish-green color. Cones oval, three to four inches long, ovate, reflexed, clustered, scales with a stout, straight, or recurved prickle. Seeds dark brown, with long, yellowish wings. Branchlets very thick, with a reddish-brown bark ; that on the old stems very thick and deeply furrowed. One of the largest and most common pines in the Rocky Mountain regions, and westward to the Pacific. Trees of these species have been found that were three hundred feet high, with stem twelve to fifteen feet in diameter, but the more usual size is from eighty to one hundred feet. The wood is quite variable, but usually it is rather coarse-grained, hard, and heavy, seldom soft, or as easily worked as the White Pine or closely allied species. I have examined and used many thousand of feet of lumber from this tree, and while admitting its value for coarse work, it is inferior as a finishing lumber to many other species. I think, however, that this tree is well adapted to dry, windy, and exposed situations, and should be tried on the western prairies, especially on light, dry, or stony soils. Several varieties are described in botanical works, but Dr. Engelmann only recognizes two, viz., var. *Jeffreyi*, a tree with a more rounded top, darker bark and paler leaves than the species. Cones also longer and lighter brown. Var. *scopulorum*, is a smaller tree, only growing about a hundred feet high, with shorter leaves, and these often in pairs. Cones only two or three inches long, grayish-brown, with stout prickles. The last variety is found throughout the Rocky Mountains, from British Columbia, to New Mexico and Arizona.

P. pungens, Michx.—Table Mountain Pine.—Leaves two in a sheath, and about two inches and a half long, rigid, stout, and of a pale yellowish-green color. Cones three inches long, ovate, sessile, usually three or four in a cluster, with woody scales, armed at the apex with a stout, slightly incurved spine on the upper scales, and recurved on the lower ones. Trees with very irregular-growing branches, and the buds covered with resin. A small tree, thirty to fifty feet high, with stem a foot or a little more in diameter. It is not a handsome or rapid-growing tree, but quite a rare one, or at least somewhat limited in its range, being found rather sparsely in Southern Pennsylvania, North Carolina and Georgia.

P. resinosa, Aiton.—Red Pine, Norway Pine.—Leaves in twos, five to six inches long, nearly cylindrical from long sheaths, rigid and straight, dark green. Cones two inches long, conical, usually in clusters, scales without points. Branchlets with reddish smooth bark. Wood hard and compact, light-colored and quite durable. A rather large tree, sixty to eighty feet high in the Eastern States, but specimens have been found in Michigan measuring a hundred and fifty feet. No large forests of this species are known, but it is found in Pennsylvania, northward to the Canadas, and west to Minnesota.

P. rigida, Miller.—Pitch Pine.—Leaves in threes, and from three to five inches long, from very short sheaths, rigid and flattened, or slightly angled on one side, of a bright, but not very dark-green color. Cones ovoid-conical, and of the size and form shown in figure 57, mostly solitary, but occasionally clustered, three or four together ; the scales terminated with a small, stout prickle. Seed small, winged. A medium sized tree, forty to seventy feet high, with stem two to three feet in diameter. Wood hard, coarse-grained, full of resin, and generally so well studded with knots as to be of little value except for fuel. A rather handsome tree when found in good soil and with room enough to grow without being crowded. More abundant in swamps and low grounds than elsewhere, but often found of

Fig. 57.—CONE OF PINUS RIGIDA.

large size on high, sandy land, slate and sandstone ridges. From Maine to Georgia, east of the Alleghanies.

Var. *Serotina*, Michx. (Pond Pine), has a little larger leaves and more ovate cones, otherwise the same as the species. Dr.

Chapman and Mr. Gordon, of England, recognize the variety as a distinct species.

P. Sabiniana, Dougl.—Sabine's Pine, Great Prickly-Coned Pine. —Leaves in threes, eight to twelve inches long, slender-drooping, of a light glaucous-green color. Cones eight to ten inches long, and four to six in diameter, of a deep mahogany-brown color, with large, projecting incurved points. Seed large, almost an inch long, sub-cylindric, with a hard, dark brown shell, and a stiff wing, only about a half inch long, with a stiff

Fig. 58.—PINUS SABINIANA.

rim. A large, round-topped tree, with thick, rough bark, and rather slender, graceful branchlets. Seed used as food by the Indians, but are not so pleasant tasted as those of *P. edulis*. A large tree, fifty to a hundred feet high, and stem two to four feet in diameter. Wood white, soft, rather even-grained, but contains a large amount of resin. Inhabits California in the Coast Ranges, and the foot hills of the Sierra Nevadas, up to an altitude of about four thousand feet. Like most of the Pines from the Coast Ranges of California, this species does not thrive in the climate of our Northern Atlantic States, but will proba-

bly thrive further South. The general form of Sabine's Pine,
when young, is shown in figure 58, taken from a cultivated species at the time the new growth is pushing out in spring.

P. Strobus, Linnæus.—White Pine, Weymouth Pine.—Leaves
five in a sheath, as shown in figure 59, and from three to four
inches long, slender, soft, and slightly whitish on the under side.
Cones from four to six inches long, cylindrical, somewhat bent
to one side, slightly drooping on rather short stalks, with
smooth, thin scales, unarmed. Seed small,
with a long wing. A well known and valuable tree, growing from one hundred to a
hundred and fifty feet high, with stem sometimes four feet in diameter. Wood white,
soft, and free from knots, and the most extensively used of any lumber in America.
But the extensive forests of White Pine,
which were to be found in our Northern
States a half century ago, are rapidly disappearing, and first-class pine lumber is already both scarce and dear. There are still
several large forests of the tree both in the
United States and the Canadas, but they will
not last long at the rate at which they are
being cut off at the present time. The White
Pine will grow rapidly on light, poor, sandy
soils, and there are millions of acres of such
lands, that could not be put to a better use
than planting it with White Pine. It is not
only a useful and handsome forest tree, but
very valuable for ornamental purposes. There
are several handsome ornamental varieties in
cultivation, the most distinct is the var. *alba*   Fig. 59.
or *nivea*, with silvery-white foliage, and var.   WHITE PINE.
nana, a dwarfish, compact little bush, with a broad, flattish head.

P. Tæda, Linn.—Loblolly Pine, Old Field Pine, Frankincense
Pine.—Leaves in threes, eight to ten inches long, from rather
long sheaths, slender, and of a light green color. Cones three
to four inches long, oblong-conical, the scales armed with a
short, rigid, straight spine. The cones are usually solitary, but
sometimes in pairs. A tree fifty to one hundred feet high, but
in some favorable situations even larger, with stem two to three
feet in diameter. Wood variable, but usually rather coarse-

grained and much inclined to warp and shrink when cut into boards and plank. A common tree in swamps, and old fields and woods throughout the Southern Atlantic States, from the southern part of Delaware, Virginia, and south to Florida, and also sparingly westward to Eastern Texas.

P. tuberculata, D. Don.—Tuberculated-Coned Pine, California Pine.—Leaves in threes, four to seven inches long, from a short, smooth sheath, slightly serrulate, and of a bright-green color. Cones three to four inches long, oblong-conical, and about two inches in diameter, in small clusters, very persistent, pendulous, of a gray color, the scales angular-tipped, with a sharp, stout prickle. A small tree, thirty to forty feet high, with stem eight to twelve inches in diameter. Wood hard, dark-colored, but too small to be of much value, except for fuel. In the Coast Ranges of California and southward.

FOREIGN SPECIES AND VARIETIES.

Of the foreign species of the Pine there are quite a large number that thrive equally as well with us as those from our own forests, and a few of them may prove even better adapted to certain soils or situations than any of our indigenous species, but this can only be determined through more extended experience with the latter. A few species of the European Pines have been quite extensively cultivated in this country for ornamental purposes, as well as for screens and wind-breaks, probably because they were to be obtained more cheaply at the nurseries than the best of our native species, but whatever the cause, the fact is quite apparent that several of the European Pines have long been favorite ornamental trees in our Atlantic States, where large and old specimens can be seen in great abundance. Nearly all the species of the Pine indigenous to the cooler region of Europe and Asia, are quite hardy in our Northern States, while those from warmer climates, including Mexico, do well in the South, but I shall only refer to a few of the best known, and to these very briefly.

P. Austriaca, Hoess.—Austrian Pine.—Leaves two in a sheath, long, slender, rigid, incurved, and sharply-pointed. Cones two to three inches long, conical, slightly recurved. Scales smooth, with a dull spine in the center. A well known and now common tree, but of comparative recent introduction, and said not to have been known in Great Britain previous to 1835, but has been raised in such immense quantities that for many years the plants could be purchased in the nurseries of Europe and in this country for a few dollars per thousand. The Austrian Pine grows to a very large size, often more than a hundred feet. Wood rather coarse-grained, but strong and moderately durable. The general habit of the tree is broad and massive, and it is of a very rapid and sturdy growth. Native of Lower Austria, Styria and adjacent regions.

P. Ayacahulte, Ehrenberg.—Mexican White Pine.—Leaves in fives, long, very slender and drooping. One of the few Mexican Pines that have proved moderately hardy in the latitude of New York. It is a large tree in its native country, growing a hundred feet high, resembling both in growth, foliage, and wood, our common White Pine, although the leaves are longer and more pendulous.

P. Cembra, Linn.—Swiss Stone Pine.—Leaves in fives, two to three inches long, very slender, triangular, straight, very numerous and crowded on the branches, and of a dark green color. Cones three inches or more in length, ovate, erect, with short but slightly hooked scales. Seeds large and nut-like, kernel edible. A very compact-growing, handsome tree, in its native country reaching a hight of a hundred and twenty feet. Native of the Alps, at elevations of four to six thousand feet, also from the Tyrol to Mount Cenis, in Austria, forming large forests. Wood resembling the White Pine of this country, and quite valuable. This species was early introduced into the United States, and has long been a favorite ornamental tree, but I regret to say that many of the oldest and finest specimens in the country have been killed by some disease, the origin of which is as yet unknown. Sometimes all the large trees in a neighborhood will die out very suddenly, the cause of this death being involved in mystery. A specimen in my grounds twenty years old, is perfectly healthy, but I have no great confidence in its longevity, for the reasons given.

Var. *Mandshuria,* Regel., is found in Japan, and of a more dwarf and compact habit than the species, otherwise scarcely distinguishable.

P. densiflora, Siebold.—Japan Pine.—Leaves in twos, about four inches long, rather large and rigid, convex above and concave beneath, very smooth, and dark, shining green ; sharp-pointed and crowded on the smaller branchlets, dropping from below when one or two years old, giving to the older branches a rather naked appearance. A common tree throughout Japan,

but most abundant in the northern and colder regions of the
country. A small tree, only thirty to forty feet high. Wood
excellent in quality, but not large enough for lumber. Hardy,
and of quite rapid growth, while young.

P. excelsa, Wallich.—Bhotan Pine.—Leaves five in a sheath,
and six or seven inches long, very slender, and of a glaucous,
green color, and very pendulous. Cones six to nine inches
long, and only about two inches in diameter, drooping and
clustered, with broad, thick, wedge-shaped imbricated scales.
One of the most graceful of all the White Pines, but very subject to blight in this country, and for this reason cannot be
recommended for general cultivation. Native of Nepal and
Bhotan on the Himalayas, at elevations of six to ten thousand
feet.

P. Laricio, Poiret.—Corsican Pine.—Leaves in twos, four to
six inches long, slender, and very wavy or somewhat twisted.
Cones two to three inches long, conical-oblong, recurved, and
of a light brown color ; scales with a minute prickle or none at
all. A large and noble tree, somewhat resembling in general
appearance the Austrian Pine, but leaves of a slightly lighter
green color, and readily distinguished by their shape. A valuable, hard, and rapid-growing tree from the South of Europe, in
the Island of Corsica, where it is said to grow to a hight of a
hundred and fifty feet. There are several varieties described in
botanical works, but none equal in value to the species.

P. Massoniana, Siebold.—Masson's Pine.—Leaves in twos, four
to six inches long, rather stiff, twisted, convex on the outer
side and concave within, quite straight, sharp-pointed, and of a
bright green color. Cones an inch to an inch and a half long,
conical, incurved, solitary, but usually very numerous, with
closely imbricated scales, terminated with slender prickles.
An upright, compact-growing tree, from forty to fifty feet
high, native of Japan, and very widely distributed from the sea-
coast to the mountains. The Japanese have several varieties of
this species in cultivation, and one known as "The Sun Ray
Pine," was introduced a few years since by the Messrs. Parsons
& Sons, Flushing, N. Y., and who have propagated it to a limited extent. The leaves of this variety are variegated with
golden-yellow, a most distinct and unique variety.

P. Mugho, Bauhin.—Mugho Pine.—Leaves in twos, one and a
half to two inches long, rigid, twisted, and of a very dark green

color. Cones small, an inch to an inch and a quarter long, and three-fourths of an inch in diameter, dark-mahogany color ; scales thin, with a triangular point, and a very minute prickle. A dwarfish tree or shrub with numerous ascending or widely-spreading branches. Quite a variable species when raised from seed, some plants assuming an erect habit, others spreading and dwarfish. My oldest specimen, twenty-five years from seed, is eight feet high and about ten feet in diameter.

P. pyrenaica, La Peyrouse.—Pyrenean Pine.—Leaves two in a sheath, and from four to seven inches long, usually crowded in tufts at the extremities of the branchlets. The color of the bark on the young growth is a bright orange color, an excellent character by which the species may be distinguished in summer. Cones two to three inches long, and about an inch and a quarter in diameter at the broadest part ; scales usually without prickles. A large tree growing sixty to eighty feet high, and native of the forests of Southern France and Spain, in the Pyrenees, mostly on the Spanish side.

P. sylvestris, Linn.—Scotch Pine.—Leaves in twos, from an inch and a half to two and a half long, twisted, quite rigid, and of a glaucous-green color, or what is sometimes called a grayish-green. Cones two to three inches long, of a grayish-brown color, with a quadrangular recurved point. Cones ripen the second year, but do not usually open until the following spring. An old and well known tree, inhabiting the colder regions of Central Europe, especially in the Tyrolian, Swiss, and Vosgian Mountains. In Europe the economic value of this tree is said to be unsurpassed by any other tree known, but the wood is not equal to our White or Southern-yellow Pine, although it is employed for similar purposes. There are a large number of varieties of the Scotch Pine, principally cultivated in Europe as ornamental trees.

CHAPTER XVIII.

ADDITIONAL LIST OF CONIFERÆ.

There are several genera of exotic conifers that are not represented in the United States by any indigenous species, and while they may never be planted here as forest trees, still quite a number have already been introduced and cultivated for ornament, and a few among them are no doubt worthy of a passing notice on this account, if not for their economic value.

AURICARIA IMBRICATA, Pavon. —*Chili Pine.*

The leaves have little or no resemblance to those of the common Pines, but are more like immense scales, from one to an inch and a half long, very broad at the base, tapering to a sharp point, and closely imbricated on the large, cane-like branchlets, which are completely covered with the dark green, and very rigid leaves. The cones are large, seven to eight inches long, nearly round, but usually a little broader than long. The seeds are large, wedge-shaped, and one to two inches long. A large tree, from one hundred to a hundred and fifty feet high, and native of the Andes of South America. Not hardy in our Northern States, but often raised in pots and boxes, and given protection in winter. Said to be perfectly hardy in England, where it is quite extensively planted for ornament.

Auricaria Cookii, Brown.—Captain Cook's Auricaria.—Leaves smaller and more slender than those of the last, and somewhat needle-shaped. Branchlets numerous and slender, the tree having quite a graceful habit. Cones three to four inches long, oval, and each scale terminated with sharp reflexed spine as shown in figure 60. A remarkable tall-growing tree, sometimes two hundred feet high, with a very slender stem. A native of New Caledonia and New Hebrides, and first discovered by Captain Cook in 1774. Quite tender even in England.

A. Cunninghamii, Aiton.—Moreton Bay Pine.—Also from Australia, where it is found—forming large forests, and growing one hundred or more feet in hight. Leaves small, stout, and very closely appressed. Cones the smallest of any species in the genus.

A. excelsa, Brown.—Norfolk Island Pine.—An enormously
large tree, some specimens having been measured that were
two hundred and twenty-five feet high, and stems eleven feet
in diameter. Tender, like all of the Australian species.

A. Rulei, Mueller.—Rule's Auricaria.—A rather small, dense,

Fig. 60.—CONE OF COOK'S AURICARIA.

and compact tree, with dark, glossy-green leaves, and large
globular cones. From Australia.

CEDRUS ATLANTICA, Manetti.—*African or Mount Atlas
Cedars.*

Leaves from one half to an inch long, almost cylindrical,
straight, rigid and sharp-pointed. Cones two to three inches

long, oval, resinous. Scales flat, smooth, and closely appressed. A large tree, somewhat of the habit of the common Larch while young, but more spreading as they become old. From the Atlas Range in Northern Africa, where it grows to a hundred feet high. Hardy in England, but not in the United States, north of Washington, but often succeeds in sheltered positions, somewhat further north.

Cedrus Deodora, Loudon.—Deodar Cedar.—A tree closely resembling the last, but with slightly longer and nearly four-angled leaves. Branches spreading and drooping. Cones four to five inches long, ovate, scales thin and closely appressed. A noble tree, from the Himalayan Mountains. It has been in cultivation for many years in this country, and at one time gave promise of being quite hardy, even in our Northern States, but now, few persons would care to risk it in any considerable number, even in the Middle States. When planted in sheltered positions, it may occasionally thrive as far north as New York, but is is not to be depended upon much north of Washington.

C. Libani, Barrelier.—Cedar of Lebanon.—Leaves about an inch long, needle-form, very much like those of the Larch, but slightly more rigid and sharper pointed. Cones similar to the last, but scales with slightly denticulate margins. A tree of great historical interest, from the mountains of Lebanon in Asia Minor, also in the mountains of Amanus and Taurus. Early introduced into England, where it appears to thrive as well as in its native country. This species is probably the hardiest of the genus, and succeeds moderately well as far north as New York in sheltered positions, and in dry, well-drained soils. There are a few old specimens of this Cedar in the suburbs of New York City, that have fruited for many years.

CRYPTOMERIA JAPONICA, Don.—*Japan Cedar.*

A genus of only this one species, which is a lofty tree in its native countries, China and Japan, where it grows to a hight of a hundred feet. Wood similar to our White Pine, and held in great esteem by the Chinese and Japanese. The leaves are small, from one half to three-fourths of an inch long, somewhat quadrangular, and sharp-pointed. Cones small, or about as long as the leaves, with numerous loose scales. There are quite a number of varieties in cultivation. A handsome, but rather uncertain tree in our Northern States, and while an occasional specimen succeeds without protection as far north as New York, it cannot be considered as hardy north of Washington.

CUNNINGHAMIA SINENSIS, R. Brown.—*Lance-Leaved Pine.*

This is another genus of only one species. A small tree, growing from thirty to forty feet high in Southern China, where it often covers the sides of the mountains, forming almost impenetrable thickets. Its leaves are from one to two inches long, flat and thin, tapering to a point. Cones an inch to an inch and a half long, oval, and mostly in clusters. Scales very small, and almost obscure, forming merely a ridge, adhering to a large, prominent, triangular bract. A very common, low-spreading shrub in nurseries and pleasure grounds, but seldom seen in good form or large enough to be classed among trees. Probably a little more hardy than the *Auricaria's*, which it resembles, but I cannot recommend it for planting out in exposed situations in any of our Northern States.

SCIADOPITYS VERTICILLATA, Siebold and Zuccarini.— *Umbrella Pine.*

A very curious and remarkable conifer, from Mount Kojasan, in the Island of Nippon, Japan, where it forms a large spreading tree, a hundred feet high. Introduced into England in 1861, and a few years later into this country. The leaves are from three to four inches long, and about one-eighth broad, double-ribbed, leathery, and blunt-pointed; dark-green, and crowded in whorls of thirty to forty at the joints or nodes of the branchlets. Cones about three inches long, and an inch and a half in diameter, solitary, with wedge-shaped corrugated, persistent scales. This curious and unique conifer gives promise of being quite hardy in our Northern States, but so few have as yet been tested in exposed situations, that a decision on this point might be considered as premature. It is a rather slow-growing tree while young, but may improve with age.

TREES NOT GENERALLY KNOWN.

Broussonetia papyrifera, Vent.—Paper Mulberry.—A rather common, small tree, in the gardens and parks of our Eastern States, and formerly quite extensively planted for ornament, but its popularity appears to be waning of late years. It is a low-growing tree, with a broad-spreading head, large, rough, ovate, or slightly heart-shaped leaves, often three cleft, or variously lobed. It is closely allied to the Osage Orange. Bark very fibrous. There are several species or varieties, all native of Japan.

Cedrela Sinensis, Juss.—Chinese Cedrela.—A strong-growing tree, native of China, with foliage resembling the ailantus, but bearing long trusses of fragrant white flowers. Its resemblance to the ailantus, led Carrière in "Revue Horticole," 1865, to give it the name of *Ailantus flavescens,* but it is more closely related to to the *Melia azedarach,* or China Tree, described on page 172. It will probably prove to be as hardy as the common Ailantus.

Cercidiphyllum Japonicum, Sieb. and Zucc.—A tall, slender-growing tree, with smooth bark, and medium sized heart-shaped leaves, of a purplish color when young, but becoming bright, glossy-green with age. Flowers very small and inconspicuous. A rare tree from Japan, and although introduced some twenty or more years ago, it has not as yet become common or even plentiful in nurseries. It is quite hardy in the neighborhood of New York City, where the oldest specimens in this country are now growing.

Eucalyptus globulus.—Fever Tree.—A large leaved, strong-scented evergreen tree, introduced from Australia, and extensively planted in California, where it has been much praised on account of its rapid growth. In its native country it is said to grow two hundred feet high, but the wood is soft, and of little value. There are an immense number of species of the *Eucalyptus,* all native of Australia, Hew Holland, and Van Diemen's Land, consequently tender in climates where there are frosts in winter. Their only merit is rapid growth and probably some slight curative properties in the balsamic odors emitted by the leaves.

Idesia polycarpa.—Maxim.—A large and handsome tree, native of Japan, with large, sub-cordate leaves, and compound racemes of diœcious, one-petaled flowers. Fruit an oranged-colored edible berry, with many seeds imbedded in a pulp. Cultivated in Japan for ornament and its edible fruit. Propagated readily from seed and cuttings of the roots. Thrives splendidly in the Southern States, but I am not fully satisfied of its hardiness in the North, although it is reported to have withstood the cold of winter in the neighborhood of Boston, while unprotected specimens have been winter-killed in the suburbs of New York City.

Phellodendron Amurense, Rupr.—Chinese Cork Tree.—A medium-sized tree from China. It is closely allied to the Prickly Ash (*Xanthoxylum*), having large pinnate leaves, which become bright red in autumn, remaining on the tree quite late. Another species is found in Japan, the *P. Japonicum.* Both species are in cultivation in this country, and are apparently quite hardy, at least they have not been injured by cold in my grounds.

Pterocarya fraxinifolia.—Spach.—A medium-sized, but rapid-growing tree from Russia and Asia. It is closely related to the Walnuts, and De Lamarck describes it under the name of *Juglans fraxinifolia* or Ash-leaved Walnut. The *Pterocaryas* are moderately hardy in our Northern States, two species having been introduced, the above and *P. stenoptera,* Cas. DC., but the latter is usually mentioned in nurserymen's catalogues under the name of *P. lœvigata.*

ADDITIONS AND CORRECTIONS.

Cercocarpus, HBK.—Mountain Mahogany.—Shrubs or small trees belonging to the *Rosaceæ* or Rose Family, only four or five species in the genus, all inhabiting the interior of North America, and only one large enough to be classed among trees.

C. ledifolius, Nutt.—Mountain Mahogany.—Leaves thick, single, evergreen, narrow lanceolate, with more or less revolute margins. Flowers small, without petals. Fruit roundish, long, hairy, included in the enlarged calyx tube. Seeds linear, with thin wings. A small tree or shrub, but sometimes thirty to fifty feet high. Wood very hard, dark mahogany-colored, rather brittle, and usually too small to be of value. Native of Oregon, Idaho, Utah, and on the slopes of the Sierra Nevada.

C. parvifolius.—Nutt.—Leaves more or less silky, and not so thick as the last, and broader or cuneate-obovate, one to one

inch and a half long, on short stalks. Flowers velvety, on short
stems. A large shrub, but sometimes twenty feet high. Wyoming Territory, Utah, New Mexico, and in the Coast Ranges
of California.

Fraxinus quadrangulata, Michx.—Blue Ash.—Leaflets five to
nine, oblong-ovate or oblong-pointed, sharply serrate, downy
beneath when young, becoming smooth when mature. Branchlets square. Seeds linear-oblong, blunt at both ends and winged
all round. A large tree, sixty to eighty feet high, with a wide
spreading top, and leaves large, sometimes eighteen inches
long. Wood similar to that of the White Ash, and excellent.
Moist, rich woods, in the Middle and Western States.

Porieria angustifolium, Gray.—A genus closely related to the
Larrea and *Guiacum*, and found along the boundary between
Mexico and the United States, from Southern Texas to California, on the dry plains. It is a small tree, with hard and
heavy wood with a brownish color. It has a local reputation as
a medicine for certain diseases of the urinal organs.

Ptelea trifoliata, Linn.—Hop-Tree.—Leaflets ovate-pointed,
downy when young. Fruit a two-celled and two-seeded
samara-winged all round, resembling an exaggerated elm seed.
They contain a bitter principle, and have been used as a substitute for hops, hence the common name. It is closely allied to
the common ailantus. Generally a large shrub, but occasionally a tree twenty-five feet high. Pennsylvania, Wisconsin, and Southward to Florida.

INDEX.

Common and Scientific names in Roman. Synonyms in *italics*.

Abele Tree 189
Abies balsamea, Marshall......... 250
   Pinus balsamea, L.
   A. balsamifera, Michx.
   Picea balsamea, Loud.
A. bracteata, Nutt.............. 250
   Pinus venusta, Dougl.
   Pinus bracteata, Don.
   Picea bracteata, Lindl.
A. cephalonica, London.......... 252
   Picea cephalonica, Loud.
A. concolor, Lindl 250
   Picea concolor, Gordon.
   Pinus concolor, Engelm.
   A. Lowiana, Murr.
   A. grandis, of California botanists.
   A. amabilis, Watson.
   Var. *A. Parsoniana*, Hort.
A. Fraseri, Lindl.......... 251
   Pinus Fraseri, Pursh.
A. grandis, Lindl.... 251
   Pinus grandis, Dougl.
   Pinus amabilis, Dougl.
   Picea grandis, Loud.
   A. Gordoniana, Carrière.
A. magnifica, Murr.............. 251
   A. amabilis, of California botanists.
A. nobilis, Lindl............... 252
   Pinus nobilis, Dougl.
   Picea nobilis, Loud.
A. Nordmanniana, Link.......... 253
   Picea Nordmanniana, Loud.
A. pectinata, DC... 253
   Picea pectinata, Loud.
   A. picea, Lindl.
   Pinus picea, Willd.
A. Pichta, Fischer.............. 253
   Abies Sibirica, Ledeb.
   Picea Pichta, Loud.
A. Pindrow, Spach.............. 253
   Picea Pindrow, Loud.
   P Herbertiana, Madd.
   P. Naptha, Knight.
A. Pinsapo, Boiss.......... 253
   Picea Pinsapo, Loud.
A. subalpina, Engelm............ 252
   A. bifolia, Murr.
   A. amabilis, Parl.
   A. lasiocarpa, Hook.
   A. grandis, of Colorado botanists.
   Var. *fallax*, Engelm. *A. amabilis*,
     Newberry.

A. Webbiana, Lindley... 253
   A. spectabilis, Spach.
   Picea Webbiana, Loud.
Acacia Greggii, Gray.............. 88
A. Three-thorned.................. 152
Acer, Maple, species of.......... .. 88
A. campestre, Linn.. 96
A. circinatum, Pursh........... 92
A. dasycarpum, Ehrhart........ .. 89
   A. eriocarpum, Michx.
   A. glaucum, M. Bieb.
A. glabrum, Torr 93
A. grandidentatum, Nutt.......... 93
A. Japonicum, Thunberg......... 98
A. Lobelii, Ten... 97
A. macrophyllum, Pursh........... 92
A. monspessulanum, Linn......... 97
A. Negundo, Linn............ 93
A. Pennsylvanicum, Linn... 92
   A. striatum, Lam.
   A. canadense, Duham.
   A. hybridum, Bosc.
A. Plantanoides, Linn........... 96
A. polymorphum, Sieb. & Zucc ... 98
A. Pseudo Platanus, Linn......... 95
   A. montanum, Lamk.
A. rubrum, Linn.................. 91
   A. Drummondii, Hook. and Arn.
A. rubrum fulgens................ 92
A. rubrum globosum..... 92
A. rubrum pyramidalis........... 92
A. rufinerve, Siebold.............. 98
A. saccharinum, Wang............. 89
A. spicatum, Lamk............... 92
   A. montanum, Ait.
A. Tartaricum, L'nn. 97
A. Tartaricum Ginnala............ 97
Æsculus, Linnæus.... 98
Æ. Californica, Nutt. 99
Æ. flava, Ait.................... 99
   Æ. Pavia flava, Mœnch.
   Æ. sargula, Buckley.
   Pavia lutea, Poir.
Æ. glabra, Willd................ 99
   Æ. Ohioensis, Michx.
   Parvia glabra, Spach.
Æ. Hippocastanum, Linn 100
Æ. parviflora, Walt............... 99
   Æ. macrostachya, Michx.
Æ. Pavia, Linn................. 100
   Æ. Paria rubra, Lamk.
Æ. rubicunda, Lois.............. 100

(285)

Æ. carnea, Willd.
Ailantus glandulosa, Desf......... 101
Alder 103
  Black... 103
  Green or Mountain............ 104
  Hoary... 103
  Oblong-leaved................. 104
  Red...... 105
  Sea-side....................... 104
  Smooth....................... 104
  Speckled 103
  White................. 104
Alerze.. 244
Algaroba...................... 190
Alligator Tree................. 164
Alnus, Tournefort... 103
A. incana, Willd... 103
  A. glauca, Michx.
  A. alpina, Bork.
A. maritima, Muhl 104
A. oblongifolia, Torr 104
A. rhombifolia, Nutt..... 104
A. rubri, Bongard............. ... 105
A. serrulata, Aiton... 104
  A. glutinosa, var. *acutifolia*, Spach.
  A. hybrida, Reich.
A. viridis, DC............. 104
  A. undulata, Willd.
  Betula crispa, Michx.
  A. fruticosa, Ledb.
Amelanchier, Medicus... 105
A. alnifolia, Nutt...... 106
  Aronia alnifolia, Nutt.
  Amelanchier florida, Lindl.
  A. canadensis, var. *alnifolia*, Torr.
  and Gray.
A. Canadensis, Torr. and Gray..... 105
  Mespilus arborea, Michx.
American cotinus 211
Amyris, Linn.................... 106
A. sylvatica, Jacq.............. 106
  A. Floridana, Nutt.
Angelica Tree................... 107
Annular Budding.............. ... 39
Anona glabra, Linn............ 111
Aralia, Linn.................... 107
A. species of.................... 107
A. spinosa, Linn................ 107
Araucaria Cookii, Brown,... 279
A. Cunninghami, Ait........... 278
A. excelsa, Brown............. . . 279
A. imbricata, Pavon........... 278
A. Rulei, Mueller.............. 279
Arbol De Hierro................ 176
Arbor Vitæ..................... 244
A. Chinese... 246
A. Columbia................... 245
A. Cutting..................... 52
A. Eastern..................... 248
A. Giant 245
A. Siberian..... 245
A. Tartarian.................... 246
Arbutus Tree, Tour.... 108
A. Menziesii, Pursh........... . 108
  A. laurifolia, Lindl.
  A. procera, Dougl.
  A. Texana, Buckley.
Arctostaphylos, Adanson.......... 108

A. Andersonii, Gray............. 109
A. bicolor, Gray.... 109
A. glauca, Lindl... 109
A. polifolia, HBK... 109
A. pumila, Nutt....... 109
A. pungens, HBK.... 109
A. tomentosa, Dougl............ 109
A. Uva-ursi, Sprengel.......... . 109
Ardisia, Swartz........ 109
A. Pickeringia, Torr. and Gray.. . 109
  Cyrilla paniculata, Nutt.
  Pickeringia paniculata, Nutt.
Arrow Wood.. 230
Ash, American................. 147
  Black.................... . . 149
  Blue..................... 283
  European.... 150
  Foreign Species and Varieties. 150
  Golden.... 150
  Green.... 149
  Oregon................... 148
  Red..................... 149
  Remilly, Weeping.... 151
  The Flowering.. 150
  Water..... 149
  Weeping..... 150
  Willow-leaved............. 150
Asimina, Adanson............. 110
A. grandiflora, Dunal.......... 110
A. parviflora, Dunal........... 110
A. pygmæa, Dunal............. 111
A. triloba, Dunal............. 109
  Anona triloba, L.
  Uvaria triloba, Torr. and Gray.
Aspen, American............. 188
  Large-toothed................ 187
Avicennia, Linn.... 111
A. nitida, Jacq............. 111
  A. tomentosa, Meyer.
  A. oblongifolia, Nutt.
Bald Cypress.................. 249
Balm of Gilead... .... 186. 250
Balsam Fir.................... 250
Balsam Tree..... 134
Bass, Sweet................... 169
Bass Wood.................. ... 225
Bastard Ironwood.............. 231
Bear Berry.............. 109, 208
Bear Berry, Buckthorn....... ... 208
Beech, American............. 145
  Antarctic................... 146
  Crested-leaved... 146
  Copper-leaved.............. 146
  Cunningham's............... 146
  Cut-leaved.................. 146
  European................... 148
  Fern-leaved................ 146
  Golden-leaved.............. 146
  Oak-leaved.. 146
  Weeping.................... 146
Best Time to Cut Timber........ 72
Betula, Tour.................. 111
B. alba, L.. var. populifolia, Spach 111
  B. acuminata, Ehrh.
  B. cuspidata, Schrad.
  B. populifolia, Ait.
B. lenta, Linn.................. 112
  B. carpinifolia, Ehrh.

B. lenta, Regel.
B. lutea, Michx... 112
  B. excelsa, Pursh.
B. nigra, Linn.. 113
  B. rubra, Michx.
B. occidentalis, Hook 113
B. papyracea, Ait................. 113
Big Tree...................... 247
Bilsted........, ... · 164
Birch, Black.................... 112
  Canoe....................... 112
  Cherry...................... 112
  Cut-leaved.................. 113
  Gray.............. ...... 111, 112
  Mahogany.................... 112
  Paper....................... 112
  River....................... 113
  Sweet...................... 112
  Weeping.................... 113
  Western..... 113
  West India.................. 114
  White....................... 111
Biota, Don.................... 245
B. Orientalis.................. 246
B. Tartarica............... 246
Bitter Wood....... 222
Black Button Wood............. 164
Black Gum........ 176
Black Haw.................. 230
Black Walnut 159
Blue Beech..................... 115
Blue Wood..................... 135
Bristly or Rose Acacia.... 215
Broussonetia papyrifera, Vent.... 282
Bois D'Arc........... 167
Bourreria Havanensis, Miers...... 113
  Ehretia Havanensis, Willd.
  B.tomentosa.var.Havanensis,Griseb.
  Ehretia tomentosa, Lam.
  Pittonia similis, Catesb.
  Ehretia Beurreria, Chapman.
  B. succulenta, Jacq.
  Var. *radula*, Gray.
  B. radula, Don.
  B. virgata, Griseb.
  Ehretia radula, Poir.
  Cordia Floridana, Nutt.
Box Elder... 93
Buckeye, Fetid 99
  Dwarf.... 99
  Ohio......................... 99
  Red.., ... 100
  Sweet... 99
Buds of Trees................. 23
Budding and Grafting............. 36
Budding Knife................... 41
Buckthorn, common............. 209
Buckthorn, southern............. 113
Buckwheat Tree........... ... x 134
Buffalo Berry................... 221
Bumelia, Swartz............... 114
B. cuneata, Swartz...... 114
  B. myrsinifolia. A.DC
  B. parvifolia. A.DC.
  B. angustifolia, Nutt.
  B. reclinata. Torr.
B. lanuginosa, Pers............... 114
  B. tomentosa, A.DC.

B. oblongifolia, Nutt.
  B. ferruginea, Nutt.
B. lycioides, Gaertn............... 114
B. tenax, Willd................... 113
Bursera, Jacquin.................. 114
B. gummifera, Jacq.............. ... 114
Butternut 159
Button Tree.... 135
Buttonwood................... 184
Buttonwood, California.......... 185
Calico-bush.................... 163
California Cedar................ 243
California Horse Chestnut........ 99
California Laurel................. 228
California Lilac.................. 129
California Nutmeg............... 234
California Olive................. 228
Cajeput.... 228
Calyptranthes, Swartz..... 114
C. Chytraculia, Swartz......... 114
  Eugenia pallens, Brown.
  Myrtus chytracula, Swartz.
Canoe Wood........ 166
Carolina Gum Tree.............. . 175
Carpinus, Linn.................. 114
C. American, Michx............ 115
Carya, Nutt.................... 115
C. alba, Nutt.................. 117
C. amara, Nutt................. 118
  Juglans, amara. Michx.
  Juglans angustifolia, Lam.
C aquatica, Nutt................. 119
  Juglans aquatica, Michx.
C. myristicaeformis, Nutt.......... 119
  Juglans myristicaeformis, Michx.
C. porcina, Nutt................ 118
  Juglans glabra, Wang.
  Juglans porcina, Michx.
  Juglans obcordata, Willd.
  C. glabra, Torr. and Gray.
C. olivaeformis, Nutt.............. 118
C. sulcata, Nutt 117
C. tomentosa, Nutt....... 117
Castanea, Tour....... 121
C. American, Michx.... 123
  C. vesca. Gaertn.
  Fagus Castanea, L.
C. Japonica..................... 125
C. pumila, Michx................. 124
  Fagus pumila, L.
C. vesca..... 123
  Castanea vulgaris, Lam.
Castanopsis, Spach............. 121
C. chrysophylla, A.DC......... 121
  Castanea chrysophylla, Hook.
  C. sempervirens, Kellogg.
Catalpa, Scopoli............ .. 125
C. bignonioides, Walt............ 125
  Bignonia Catalpa, L.
  C. cordifolia, Jeanne.
  C. syringaefolia Sims.
  C. speciosa. Warder.
C. Bungei, C. A. Mey.... 128
C. common..... 125
C. Golden..................... 128
C. Japan.... 128
C. Kaempferi. DC................. 128
  C. Japonica.

C. ovata, Geo. Don.
C. Speciosa, Warder... 125
Cat's-Claw................... 183
Ceanothus, Linn. 128
C. spinosus, Nutt 129
C. thyrsiflorus, Eschscholtz....... 124
Cedar, Deodar................... 280
    Japan...................... 282
    Lebanon ....•....... 280
    Mount Atlas.................. 279
    Red...................... 239
    White.. 242
Cedrela sinensis, A. Juss 282
C. Atlantica, Manetti........... .. 279
C. Deodora, Loud.............. ... 280
C. Libani. Bour............. 280
Celtis, Tour................. ... 129
C. brevipes, Watson... 129
C. Mississippiensis, Bosc. 129
    C. occidentalis, var. *tenuifolia*, Pers.
    C. lævigata, Willd.
    C. occidentalis, var. *integrifolia*,
      Nutt.
    C. integrifolia, Nutt.
    C. longifolia, Nutt.
C. occidentalis, Linn.............. 129
    C. crassifolia, Lam.
    C. occidentalis, var. *crassifolia*, Gray.
C. Tala, Gillies, var. pallida. Planch 131
    C. (Momisia) pallida, Torr.
Cercidiphyllum Japonicum........ 282
Cercis, Linn....... 131
C. Canadensis, Linn./.............. 131
C. Japonica, Siebold. 132
    C. Chinensis, Bunge.
C. occidentalis, Torr... 131
    C. Californicum, Torr.
C. siliquastrum, Linn. 131
Cercocarpus, HBK.... 283
C. ledifolius, Nutt.... 283
C. purvifolius, Nutt................ 283
Chamæcyparis, Spach.... 241
C. Lawsoniana. Parl.... 241
    Cupressus Lawsoniana, Murr.
    Cupressus Nutkaensis, Torr.
    Cupressus fragrans, Kellogg.
    Cupressus attenuata, Gordon.
C. Nutkaensis, Lam... 241
    Thuya excelsa, Bong.
    Cupressus Nutkaensis, Lamb.
    Cupressus Americana, Trautv.
    C. excelsa, Fisch.
    Thuyopsis borealis, Hort.
    Thuyopsis Tchugatskoy, Hort.
C. thujoides, Lam............... 242
    Thuya sphæroidalis, Rich.
C. torulosa, Don............ 243
Characteristics of Trees, The...... 19
Cherry, Anderson's... 192
    California...................... 193
    Dwarf or Sand.................. 195
    Holly-leaved.................. 192
    Laurel... 192
    Wild, Black.................. 195
    Wild, Red.... 192
Chestnut, American..... 123
    California.................... 121
    Chinquapin. 124

Cut-leaved..................... 125
    Dwarf....................... 124
    European... 121
    Golden....................... 121
    Japan....................... 125
    Numbo..................... 123
Chilopsis saligna, Don........... 132
    C. linearis, DC.
    Bignonia linearis, Cav.
    C. glutinosa, Engelm.
China Tree....................... 172
Chinquapin...................... 124
Chionanthus Virginica, Linn. .. . 132
C. var. angustifolia.............. 133
Choke Berry.................... 182
Chrysophyllum microphyllum, DC. 133
C. oliviforme, Lam............. . 133
    C. monopyrenum, Swartz.
Cladrastis tinctoria, Raf......... 133
    Virgilia lutea, Michx.
C. Amurensis, Benth. and Hook... 134
    Maackia amurensis, Rupr.
Clammy Locust... 215
Cliftonia ligustrina, Banks........ 134
    Mylocarium ligustrinum, Willd.
Clusia flava, Linn.... 134
Coccoloba, Jacq... 135
C. Floridana, Meisner............ 135
    C. parvifolia, Nutt.
C. unifera, Jacq...... 135
Coffee Tree, Kentucky............ 154
Condalia, Cavan................. 135
C. obovata, Hook............... . 135
Coniferæ 237
Cone-bearing Trees.............. 237
Coniferæ from Cuttings........... 51
Conocarpus, Linn............... 135
C. erecta, Jacqu................. 135
Coral Sumach.................. 214
Cordia, Linn—Plumier............ 136
C. Boissieri, DC................. 136
C. Sebestena, L................. 136
    C. speciosa, Willd.
Cornus, Tour 136
C. florida, Linn 136
C. Nuttallii, Audubon.... 136
Cotton Gum....................... 176
Cottonwood..................... 187
Crab Apple, American 182
    Narrow-leaved................ 181
    Oregon....... 182
Crab Wood...................... 220
Cranberry Tree................. 230
Cratægus, Linn................. 138
C. æstivalis, Torr. & Gray........ 138
C. apiifolia, Michx.... 139
C. arborescens, Elliott.......... 139
C. berberifolia. Torr. & Gray...... 139
C. coccinea. Linn 139
C. cordata, Ait................. 139
C. Crus-galli 139
C. Douglasii. Lindl............. 139
    C. sanguinea, var. *Douglasii*,
      Torr. & Gray................. 139
C. flava. Ait................... 139
C. parviflora, Ait.............. 140
C Pyracantha.................... 141
C. rivularis, Nutt. 140

C. spathulata, Michx............. 140
   C. microcarpa, Lindl.
C. subvillosa, Schrad........ 140
   C. coccinea, var. *mollis*, Torr. &
   Gray.
   C. tomentosa, var. *mollis*, Gray.
   C. mollis, Scheele.
C. tomentosa, L. 140
Cryptomeria Japonica, Don...... 280
Cucumber Tree.................. 168
Cunninghami Sinensis, R. Brown. 281
Cupressus, Tour..... 240
C. Goveniana, Gordon.......... 240
C. Macnabiana, Muir........... 240
C. macrocarpa, Hartw............ 240
   C. lambertiana, Gord.
   C. Hartwegii, Carrière.
Custard Apple.................. 110
Cutting of Cypress............. 53
Cypress, Bald.................. 248
C California.................... 240
C. deciduous................... 249
C. Lawson's.................... 241
C. McNab's 240
C. Monterey.................... 240
C. Nootka Sound................ 241
C. Twisted-branched............ 242
C. Weeping..................... 249
Cyrilla, Linn... 141
C. racemiflora, Walt. 141
   C. Caroliniana, Richard.
Dacridium..................... 236
Dahoon Holly..... 157
Deciduous Cypress.............. 249
Deciduous Trees from Cuttings... 54
Desert Willow.................. 132
Devil Wood.................... 176
Diospyrus, Linn................ 142
D. Texana, Scheele............. 142
D. Virginiana, L............... 142
Dogwood, Flowering............. 137
D. Nuttall's................... 137
D. Weeping.................... 137
D. Western Species............. 137
Dipholis salicifolia, A. DC 143
   Achras salicifolia, L.
   Bumelia salicifolia, Swartz.
Drypetes. Vahl................. 143
D. crocea, Poit................ 143
   Schæfferia lateriflora, Sw.
Ebretia, Linn........ 143
E. elliptica, DC............... 144
Elder Tree.................... 218
E. Black-berried.............. 219
E. European................... 219
Elm, American................. 227
E. Dutch Cork-bark............ 228
E. Exeter.... 229
E. English.................... 228
E. English Cork-bark.... 228
E. Opaque-leaved.............. 227
E. Red 227
E. Slippery................... 227
E. Scotch 228
E. Small-leaved............... 227
E. Weeping................ ... 228
E. Wych...................... 228
Encino........................ 147

Establishing New Forests........ 80
Eucalyptus globulus............ 282
Eugenia, Micheli.............. 144
E. Box-leaved................. 144
E. buxifolia, Willd. 144
   Myrtus buxifolia, Swartz.
   M. axillaris, Poiret.
E. dichotoma, DC............. 144
   Myrtus dichotoma, Vahl.
   Eugenia fragrans, Willd.
   E. montana, Aubl.
   E. divaricata, Lam.
E. procera, Poir.............. 144
   Myrtus procera, Swartz.
E. Small-leaved............... 144
E. Tall...................... 144
Evergreens from the Forests...... 64
Fagus, Tour 144
F. antarctica, Forst. 146
F. betuloides, Mirb 146
F. Cunninghami, Hook........ 146
F. ferruginea, Ait 145
   Fagus sylvestris, Michx.
   F. Sylvatica Americana, Loud.
   F. alba, Rafinesque.
F. sylvatica, Linn 145
   Castanea Fagus, Scop.
False Acacia......... 215
   Box........................ 220
   Elm........................ 129
Fever Tree....... 282
Ficus, Tour 146
F. aurea, Nutt............... 146
F. brevifolia, Nutt 146
F. pedunculata, Ait 147
Fig, Cherry.................. 147
F. Short-leaved.............. 146
F. Small-fruited... 146
Fir, Balsam..... 250
   Black Balsam............... 250
   Cephalonian................ 252
   Cicilian Silver............. 252
   European Silver............ 253
   Fraser's Balsam..... 251
   Great Silver............... 251
   Leafy-bracted Silver........ 250
   Noble Silver............... 252
   Nordmann's.. 253
   Pinsapo.................... 253
   Upright Indian............. 253
   Red........................ 251
   Siberian Silver............ 253
   Webb's Purple-coned........ 253
   White..................... 250
Florida Myrtle................ 174
Forest Trees, Description of...... 87
Forests and Insects.. 18
Forests and Streams.. 15
Forked Calyptranthes.... 114
Fraxinus, Tour 147
F. Americana, Lim............ 147
   F. acuminata, Lam.
   F. alba, Marsh.
   F. juglandifolia, Lam.
   F. epiptera, Michx.
   F. Curtissii, Vasey.
F. anomala, Torr............. 148
F. cuspidata, Torr............ 148

F. dipetala, Hook. & Arn......... 148
   Ornus dipetala, Nutt.
F. Greggii, Gray...... 148
F. Oregona, Nutt................ 148
   F. grandifolia, Benth.
   F. pubescens, var., Hook.
F. pistaciæfolia, Torr............. 148
   F. velutina, Torr.
   F. coriacea, Watson.
F. platycarpa, Michx.............. 149
   F. Caroliniana, Lam.
   F. Americana, Marsh.
   F. pallida, Bosc.
   F. pauciflora, Nutt.
   F. triptera. Nutt.
F. pubescens. Lam... 149
   F. Pennsylvanica, Marsh.
   F. nigra, DuRoi.
   F. tomentosa, Michx.
F. quadrangulata, Michx......... 283
   F. tetagona, Cels.
F. sambucifolia, Lam......... 149
F. viridis, Michx. 149
   F. concolor, Muhl.
   F. juglandifolia, Willd.
   F. Caroliniana, Willd.
   F. expansa, Willd.
   F. Berlandieriana, DC.
Fringe Tree, White.............. 132
Genip Tree... 157
Georgia Bark 180
Ginkgo 236
Gleditschia. Linn................. 152
G. Carpica, Desf 153
G. monosperma, Nutt........... . 153
G. Sinensis, Lamk................ 153
G. triacanthos, L................. 152
G. var. Bujoti pendula........... 152
G. var. inermis.................. 152
Glyptostrobus. Endl. 249
G. heterophyllus, Endl........... 249
G. pendulus. Endl................ 249
Gordonia, Ellis.... 153
G. Lasianthus. L................. 153
G. pubescens, L'Her............. 153
Great Laurel............ 204
Great Tree of California 247
Grafting Cleft.................. ... 45
G. Conifers...................... 48
G. Crown........... 46
G. Deciduous Trees.............. 42
G. Pine 49
G. Splice or Tongue............. 47
G. Terminal. 50
G. Side or Triangular........... 47
G. Wax 43
Guaicum, Plumier............... 154
G. sanctum, L................... 154
Guilder Rose. 231
Gymnocladus, Lam.............. 154
G. Canadensis, Lam..... 154
Hackberry...................... 129
Hackmatac... 259
Halesia, Ellis... 155
H. diptera, Linn................ 155
H. Four-winged. 155
H. parviflora, Michx............ 155
H. tetraptera, L.... 155

H. Small-flowered.... 155
H. Two-winged 155
Hemlock, California... 255
   Canada................. 254
   Patton's 255
Hercules' Club.... 107
Heteromeles arbutifolia, Rœmer.. 156
   Aronia arbutifolia. Nutt.
   Cratægus arbutifolia. Poir.
   Photinia arbutifolia, Lindl.
   Mespilus arbutifolia, Link.
   Photinia salicifolia. Presl.
   H. Fremontiana Dcsne.
Hickory. Bitter Nut........ 118
   Brown...... 118
   Hales, Paper shell............. 119
   Nutmeg...................... 119
   Mocker Nut... 117
   Pecan Nut................. 118
   Pig Nut....... 118
   Shag-bark............. 117
Shell-bark..................... 117
Swamp....... 118
Thick Shell-bark............. .. 117
Western Shell-bark.............. 117
   White-heart............... 117
Hippomane Mancinella.. 156
Hog-plum..................... 232
Holly, American................ 157
   Dahoon...................... 157
   European.................... 157
   Yaupon..................... 158
Honey Berry.............. 156
   Locust...................... 152
   Mesquit.................... 190
Hop-hornbeam. American........ 177
Hornbeam, American........... 215
   European... 177
Horse-chestnut, Cut-leaved....... 100
   Double White........ 100
   European.................. 100
   Memminger's......... 100
Horse Sugar.................... 224
Hypelate paniculata, Cambess..... 57
   Melicocca paniculata, Juss.
H. trifoliata, Swartz.............. 157
Idesia. polycarpa, Maxim........ 282
Ilex. Linn..................... . 157
I. Cassine, Linn...... 158
I. cor acea, Ell 158
I. Dahoon, Nutt............... 157
   I. ligustrina, Ell.
   I. laurifolia. Nutt.
   I. myrtifolia, Walt.
I. glabra, Gray................. 158
I. opaca, Ait................... 157
I. verticillata. Gray 158
Influence of Forests on Climate... 9
Ink-berry 158
Implements Used in Pruning 71
Importance of a Supply of Wood. 75
Inga Unguis, Catl.............. 183
Iron Wood........ .. 115, 122, 176, 177
Jamaica Dogwood... 182
Judas Tree..................... 131
   California.... 131
   European.................... 131
   Japan...................... 132

Juglans, Linn...................159
J. Californica, Watson..........159
   J. rupestris, var. *major*, Torr.
J. cinerea, L....................... 159
   J. oblonga, Mill.
   J. cathartica, Michx.
J. nigra, L.. 159
J. regia. Linn 160
J. rupestris, Engelm.. 160
Jujube...................... ...232
June Berry..................... 105
Juniper....................... 238
J. California.... 239
J. Common...................236
J. Western. ... .239
Juniperus, Linn... 238
J. Californica. Carrière. 238
   J. tetragona, var. *osteosperma*,
   Torr.
   J. Cerrosianus, Kellogg.
   J. occidentalis, Parl.
   Var. *Utahensis*, Engelm.
   J. occidentalis, Watson.
   J andina. Nutt.
J. communis, Linn 239
   J. depressa, Pursh.
   J. Canadensis, Ladd.
J. occidentalis, Hook............ 239
   J. excelsa, Pursh.
   Var. *conjungens*, Engelm.
   Var. *monosperma*, Engelm.
J. pachyphlœa. Torr.. 239
   J. plochyderma, Torr.
J. Virginiana, Linn....... 239
Kalmia. Linn 163
K. angustifolia. Linn ....: 163
K. Broad-leaved. ... .163
K. cuneata, Michx................163
K. glauca, Ait 163
K. Hairy-leaved.... 163
K. hirsuta, Walt 163
K. latifolia, Linn.................163
K. Narrow-leaved....:. 163
Kentucky Coffee Tree.............154
Kiaka Elm...................184
Kinnikinick..... 109
Ladder, A Handy 71
Laguncularia racemosa, Gœrtn......164
Larch, American..................259
L. Dahurian...................260
L. Golden 261
L. Japan........ 261
L. Sikkim..................261
Larix, Tour. 259
L. Americana, Michx 259
   Pinus pendula. Ait.
   L. pendula, Salisb.
   L. macrocarpa. Forbes.
   L. intermedia. Lodd.
   Pinus microcarpa. Lamb.
L. Dahurica, Turz...............260
L. Griffithiana. Hook.. 261
L. Kæmpferi. Gordon............261
L. Leptolepsis. Gordon.............261
L. Lyallii, Parl. 239
   Pinus Lyallii, Parl.
L. occidentalis, Nutt 260
   L. Americana, var. *brevifolia*, Car.

   Pinus Nuttallii, Parl.
Lath-covered Frame.............. 59
Laurel, American.............. 163
   Carolina..................163
   Pale... 163
   Sheep....................163
Layers....... 54
Layer in a Pot.... 57
Layering a Branch............... 56
Lever-wood.................. 177
Libocedrus, Endl 243
L. decurrens, Torr..............243
   Thuya Craigiana, Balfour.
   Thuya gigantea, Carrière.
   Heyderia decurrens, Koch.
L. tetragona, Endl.. 244
Lignum Vitæ 154
Lilac, California.............. .. 129
Linden Tree....................225
Liquidambar Styraciflua, Linn.....164
L. imberbe, Ait................166
L. Longworthii, Thurber........165
L. orientalis. Mill.............166
   Platanus orientalis, Pocke.
Liriodendron Tulipifera, Linn......166
Loblolly Bay...................153
Locust.......................214
   Tree....................215
   Carpian Honey........153
   Chinese Honey...........153
   Honey..................152
   Water...................153
Logwood.....................155
Lombardy Poplar...............189
Maclura aurantiaca, Nutt.. 167
Madeira Wood...............157
Madrono.... 108
Magnolia, Linn...............167
M. acuminata, Linn............ 168
M. cordata, Michx 168
M. Fraseri, Walt.... 168
   M. arriculata, Lam.
   M. pyrimidata, Bartram.
M. glauca. L................. 169
M. grandiflora, L.............. 16
M. macrophylla, Michx..........169
M. Umbrella, Lam 170
   M. tripetala, L.
M. Thompsoniana...... 170
Magnolia, Chinese Species and
   Varieties...... 170
M. atropurpurea.................. 170
M. conspicua..................170
M. hypoleuca................. 171
M. Kobus...................... 171
M. Lennei.................... 171
M. Norbertiana................ 171
M. parviflora................ 171
M. purpurea................. 171
M. speciosa................. 171
M. Soulangeana............. 171
M. stellata................ 171
M. stricta................. 171
M. superba....:........... 171
Magnolia, Chinese White......... 170
   Chinese Superb.............. 170
   Ear-leaved................. 168

Great Chinese................. 171
Great-leaved.................. 169
Hall's Japan..... 171
Japan Purple................. 171
Large-flowered...... 169
Lenne's Hybrid..... ... 171
Norbert's.................... 171
Showy....................... 171
Small-flowered....... 171
Soulange's Hybrid............ 171
Star......... 171
Swamp...................... 169
Thompson's.................. 170
Mahogany Tree................. 224
Maiden Hair Tree.............. 236
Management of Forests.......... 79
Manchineel..... 156
Mangrove 209
Manzanita.................... 108
Manzanita, California.. 109
Maple......................... 88
  Ash-leaved...... 93
  Black....................... 89
  California................... 92
  Crisp-leaved................ 90
  Cut-leaved Norway..... 96
  Eagle's-claw................. 96
  English Field............... 96
  European.................... 95
  Foreign Species of........... 95
  Ginnala 97
  Golden-leaved.............. 95
  Hard..... 89
  Japan........ 97
  Large-leaved... 92
  Lobel's.... 97
  Lorberg's................... 96
  Montpelier................. 97
  Mountain.. 92
  Mountain Sugar..... 93
  Negundo................... 93
  Norway... 96
  Purple-leaved.............. 95
  Red.............. 91
  Rock....................... 89
  Round-leaved............... 92
  Schwerdler's Norway.......... 96
  Silver...................... 89
  Silver Striped.............. 95
  Smooth-leaved Mountain..... 93
  Striped-bark.. 92
  Sugar...................... 89
  Swamp...... 91
  Tartarian................... 97
  Three-colored leaved........ 95
  Three-lobed................ 97
  Velvet-leaved............. ... 95
  Vine.. 92
  Wagner's cut-leaved. 90
  Weir's cut-leaved............ 90
  White. 89
Melia, Linn................... 171
M. Azedrach, Cav............. 172
Mesquit 189
Mesquit Tree................. 191
Mimusops, Linn 172
M. Sieberi, A.DC.............. 172
  M. dissecta, Griseb.

Acras Zapotilla, var. *parviflora*,
    Nutt
Mock Orange................. 192
Moose Wood.... 92
Morus, Tour 172
M. alba, Linn 173
M. microphylla, Buckley.......... 173
M. rubra, L........... 172
  M. Canadensis, Lam.
Mountain Ash, American.......... 181
Mountain Ash, Western... 182
Mountain Mahogany............. 28
Mountain Manchineel.......... 156
Movement of Sap in Trees........ 52
Mulberry, Downing's............ 173
  Red....................... 172
  Russian 173
  Tartarian................... 173
  West Indian................ 173
  White................. 173
Myrica, Linn................. 174
M. Californica, Cham.......... 174
Myrsine, Linn................ 174
M. Rapanea, Roem, and Schult.. ... 174
  M. floribunda, Griseb.
  M. Floridana, A.DC.
  Rapanea Guyanensis, Aubl.
  Semara floribunda, Willd.
Naseberry................... 172
Negundo aceroides, Moench....... 93
N. Californicum, Torr. and Gray... 94
Nettle Tree... 129
Nettle Tree, Southern... 129
New Forests, Establishing of...... 80
Northern Prickly Ash............ 232
Nuttallia cerasiformis, Torr. and
    Gray.................... 174
Nyssa, Linn 175
N. capitata, Walt............ 172
  N. candicans, Michx.
N. Caroliniana, Poir 175
  N. aquatica.
N. multiflora, Wang.... 175
  N. aquatica. L.
  N. biflora, Michx.
N. sylvatica. Marsh... 176
  N. villosa. Michx.
  N. multiflora, var. *sylvatica,* Watson.
N. uniflora, Wang.............. 176
  N. aquatica, L
  N. tomentosa. Michx.
  N. grandidentata, Michx
Oak, Barren............... 204
  Bartram....... 201
  Bear... 202
  Black....................... 205
  Black Jack................... 204
  Black Scrub... 202
  Blue....................... 200
  Brewer's... 198
  Burr....................... 203
  California Chestnut.... 203
  California Live.............. 199
  Chinquapin 204
  Cut-leaved................. 207
  Daimio 207
  Dwarf Evergreen... 201

Evergreen, White............ 201
Georgia.................. 201
Golden-leaved................ 207
Holly-leaved.... 197
Hybrids 207
Kellogg's...... 202
Laurel-leaved........ 202
Live 206
Lobed-leaved.............. 203
Mossy-cup 203
Mottled-leaved..... 207
Mountain White............. 200
Ornamental 207
Over-cup...... 203
Palmer's.. 204
Post.. .... 203, 204
Quercitron ....-..... 204
Red........ 205
Rocky Mountain Scrub....... 206
Scarlet................... 199
Scrub................... .. 199
Shingle..... 202
Small-leaved.. 200
Spanish............. 201
Swamp........... 204
Swamp Chestnut... 204
Swamp, White.... 198
Turkey................. ... 199, 207
Water................... 198
Western................... ... 201
White............ 197
Willow............ 204
Yellow-barked............... 204
Yellow Chestnut 504
Ogeechee Lime................... 175
Olea, Linn...... 176
Olneya Tesota, Gray......... 176
Osage Orange............. 167
Osier.......... 215
Osmanthus Americanus, Benth. and
     Hook.. 176
     Olea Americana, L.
Oso Berry...................... 174
Ostrya, Michell 177
O. Virginica, Willd.......... 177
     Carpinus Ostrya, L.
     Carpinus Virginiana, Lam.
     O. Americana, Michx.
     O. vulgaris, Watson.
     Carpinus triflora, Mœnch.
Oxydendrum arboreum, DC.. 179
     Andromeda arborea, L.
Palo verde...... 180
Papaw............................ 110
     Dwarf..... 111
     Large-flowered. 110
     Small-flowered..... 110
Parkinsonia, Linn.... 179
P. aculeata, Linn................ 179
P. florida, Watson.......... 179
     Cercidium floridum, Benth.
P. macrophylla, Torr............ 179
P. Torreyana, Watson............ 179,
     Cercidium floridum, Torr.
Paulownia, Siebold.......... 180
P. imperialis, Siebold... 180
     Bignonia tomentosa, Thunb.
Pepperidge...................... 175

Persea, Gærtn........... 180
P. Carolinensis, Nees.......... 180
     Laurus Borbonica, L.
     Laurus Carolinensis, Catesb.
     P. Borbonica, Spr.
P. Catesbyana, Michx.. 180
     Laurus Catesbyana, Michx.
Persimmon, common..:........ ... 142
Persimmon, Mexican....... 142
Phellodendron amurense, Rupr.... 283
Picea alba, Michx............. ... 256
     Pinus alba, Ait.
     Abies alba, Michx.
P. Engelmanni, Parry............. 256
     Abies nigra, Engelm.
     Abies Engelmanni, Parry.
     Pinus commutata, Parl.
P. excelsa. DC........... 257
P. firma, Gord........... 258
P. nigra, Poiret...... 256
     Pinus nigra, Ait.
     Abies nigra, Michx.
     Pinus rubra, Lamb.
     Abies rubra, Poir.
     Abies nigra, Michx.
     P. rubra, Link.
P. orientalis, Poiret............ 258
P. polita, Sieb. and Zucc.......... 258
P. pungens, Engelm............. 257
     Abies Menziesii of Colorado botan-
          ists.
P. Sitchensis, Bongard 257
     Pinus Sitchensis, Bong.
     Pinus Menziesii, Dougl.
     Abies Menziesii, Lindl.
Pigeon Plum..................... 135
Pinckneya pubens, Michx........ 180
Pine, Austrian 274
     Bhotan 276
     Cat-tail...................... .. 263
     Chihuahua 264
     Chili 278
     Corsican....... 276
     Coulter's...................... 265
     Elliott's...................... 266
     Fox-tail...................... . 263
     Frankincense................. 273
     Fremont's 268
     Georgia....................: 263
     Gray....................... 254
     Great Prickly-coned 272
     Heavy Wooded................ 270
     Hickory................... 263
     Hooked-cone 265
     Japan.............. 275
     Jersey....................... 267
     Lambert's..................... 268
     Lance-leaved.... 281
     Loblolly........ 273
     Long-leaved 263
     Mandshurian................. 275
     Masson's...................... 276
     Mexican White................ 275
     Monterey................... 267
     Moreton Bay................. 278
     Mountain.................. 269
     Mugho 276
     Norfolk Island................. 279

Norway............................... 271
Nut.................. .... 266, 268
Old Field......................... 273
Parry's................. 269
Pinon................... 266
Pitch................... 271
Pond....................... 271
Pyrenian....................... 277
Red Pine 271
Sabine's 272
Scotch................... 277
Scrub...................... 264, 267
Short-leaved 268
Southern Yellow........... 263
Spruce...................... 267
Sugar....................... 268
Swiss Stone............... 275
Table Mountain.............. 270
Tuberculated-coned 274
Twisted Branched.......... 264
Umbrella... 277
Western White.... 266
Weymouth................... 273
White 273
Yellow............. 263, 268. 270
Pinus, Tour....... 262
P. Arizonica, Engelm 263
P. australis, Michx 263
  P. palustris, Mill.
  Var. excelsa, Loud.
  P. palustris excelsa, Booth.
P. Austriaca, Hoess 274
  P. nigra, Link.
  Laricio Austriaca, Endl.
P. Ayacahuite..................... 275
P. Balfouriana, Jeffrey... 263
  Var. aristata, Engelm.
  P. aristata, Engelm.
P. Banksiana, Lamb.............. 264
  P. Hudsonica, Poir.
  P. rupestris, Michx.
P. cembra, Linn................... 275
  Var. Mandshuria, Regel.
P. Chihuahuana, Engelm.. 264
P. contorta. Dougl.. 264
  P. inops, Boug.
  P. Bolanderi, Parl.
  Var. Murrayana, Engelm.
  P. contorta, Newberry.
  P. inops, Benth.
  P. contorta, var. latifolia, Engelm.
  Murrayana, Murr.
P. Coulteri, Don.............. 265
  P. macrocarpa, Lindl.
P. densiflora, Siebold.............. 275
  P. Japonica, Antoine.
  P. Pinea, Gordon.
P. edulis, Engelm.................. 266
  P. embroides, Gordon.
P. Elliottii, Engelm... 266
P. excelsa, Wall.................... 276
  Strobus excelsa, Gordon.
P. flexilis, James................ 266
  Var. albicaulis, Engelm.
  P. cembroides, Newberry.
  P. albicaulis. Engelm.
  P. Shasta. Carrière.
P. glabra, Walt...................... 267

P. inops, Ait....................... 267
P. insignis, Dougl........ 267
  P. Californica, Lois.
  P adunca, Bosc.
  P. radiata, Don.
  P. tuberculata, Don.
P. Lambertiana. Dougl........ 268
  P. Strobus Lambertiana, Gord.
P. Laricio. Poiret........... 276
P. Massoniana, Sieb............... 276
  P. sylvestris, Thunberg.
  P. rubra, Siebold.
  P. Pinaster, Loud.
P. mitis, Michx.................... 268
  P. variabilis, Pursh.
P. monophylla, Torr. and Frem.... 268
  P. Fremontiana, Endl.
P. monticola, Dougl............... 269
  P. Strobus Monticola, Loud.
P. Mugho, Bauhin.................. 276
  P. Mughus. Loud.
  P. Sylvestris, Mugho Bauhin.
  Var. rostrata, Antoine.
  Var. rotundata, Link.
P. muricata, Don..... 269
  P. Murrayana, Balfour.
  P. Edgariana, Hartw.
P. Parryana, Engelm.............. 269
  P. Llaveana, Torr.
P. Ponderosa, Dougl.............. 270
  P. Benthamiana, Hartw.
  P. Beardsleyi, Murr.
  P. Craigana, Murr.
  Var. scopulorum, Engelm.
  Var. Jeffreyi. Engelm.
P. pungens, Michx................ 270
P. pyrenaica, Le Pey. 277
  P. Hispanica, Cook.
  P. penicellus, Le Pey.
  P. Laricio Pyrenaica, Loud.
P. re-inosa, Ait.. 271
  P. rubra. Michx.
P. rigida, Mill.. 271
  Var. Serotina, Michx.
P. Sabiniana, Dougl... 272
P. Strobus, Linn............... ... 273
P. sylvestris, Linn 277
P. Tæda, Linn..................... 273
P. tuberculata, D. Don............. 274
  P. Californica, Hartw.
Pirus........... 181
P. Americana, DC................. 181
  Sorbus Americana, Marsh.
P. angustifolia, Ait. 181
  Malus angustifolia, Michx.
P. arbutifolia, Linn.... 182
  Aronia arbutifolia, Ell.
P. coronaria, L.................... 182
  Malus coronaria, Mill.
P. rivularis, Dougl................ 182
  Malus rivularis, Deane.
  Pirus diversifolia. Bongard.
P. sambucifolia, Cham. and Schlect 182
  Sorbus sambucifolia, Rœm.
Piscidia Erythrina, Linn.......... 182
Pistacia, Will. 183
P. Mexicana, HBK................ 183
P. Nut.............................. 183

Pithecolobium, Martin............ 183
P. Unguis-Cati, Benth... 183
   Inga Unguis-Cati, Willd.
   P. Guadalupense, Nutt.
Plane Tree, American............. 185
Plane Tree, Oriental............. 185
Planera aquatica, Gmel............ 183
   P. Gmelini, L. C. Rich.
   P. ulmifolia, Michx.
   Anonymos aquatica, Walt.
Planer Tree........... 183
Planer Tree. Caucasian...... 184
Plantanus, Tour 184
P. acerfolia...... 185
P. asplenifolia.... 185
P. lirlodendrifolia.............. 185
P. Orientalis, Linn............... 185
P. occidentalis, L........ 184
P. quinquelobata.... 185
P. racemosa, Nutt............... 185
P. Wrightii, Watson............. 185
Plum........... 192
   Beach...................... 193
   California 195
   Chickasaw... 192
   Evergreen... 193
   Wild..... 192
Podocarpus Japonica........... 236
Podocarpus, South American.... 236
Poison Dogwood................ 211
   Ivy..................... ... 211
   Oak..................... ... 210, 211
   Sumach 211
   Wood...... 220
Poplar. Balsam.................. 186
   Black...................... 189
   Carolina....... 187
   Crisp-leaved or curled-leaved.. 189
   Downy-leaved.............. 187
   Lombardy.................. 189
   Silver.................... 189
   Weeping.................. 189
   Willow-leaved.... 186
Populus, Tour..... 185
P. alba, Linn............... 189
P. angustifolia, James............. 186
   P. Canadensis, var. *angustifolia,*
   Wesmael.
   P. balsamifera, var. *angustifolia,*
   Watson.
P. balsamifera, Linn.... 186
   Var. *candicans,* Gray,
P. dilatata, Tour............. 189
P. fastigiata, Desf....... 189
   P. nigra. Cat-sb.
   P. macrophylla, Lindl.
   P. Ontariensis. Desf.
   P. suaveolens, Fischer.
P. Fremontii. Watson 187
   P. monilifera, Newberry.
   P. monilifera.
   Var. *Wislizeni.* Watson.
P. grandidentata, Michx......... 187
P. heterophylla, Linn............. 187
   P. argentea, Michx.
   P. heterophylla, var. *argentea,* Wes-
   mael.
   P. cordifolia, Burgsd.

P. monilifera, Ait.... 187
   P. angulata, Ait.
   P. angulata, Michx.
   P. Canadensis, Desf.
   P. Marylandica, Bosc.
   P. lævigata, Willd.
   P. glandulosa, Mœnch.
P. nigra, Linn.................... 189
P. tremuloides, Michx.............. 188
   P. græca, Willd.
   P. benzoifera, Tausch.
P. suaveolens, Regel 189
P. trichocarpa, Torr. and Gray..... 188
   P. balsamifera, var. Hook.
   P. b ilsamifera, var. Watson.
Porliera angustifolia. Gray........ 284
   Guiacum angustifolium, Engelm.
Preparing a Seed-bed............. 28
Preservation of Forests........... 78
Prickly Ash....... 231
Pride of India.................. 172
Propagation by Layering.......... 55
Prosopis. Linn 189
P. juliflora, DC 190
   Algarobia glandulosa, Torr. & Gray.
P. pubescens, Benth 190
   Strombocarpa pubescens, Gray.
Pruning Evergreens............. 70
Pruning of Forest Trees........... 67
Prunus Americana, Marshall....... 192
P. Andersoni, Gray............... 192
P. Caroliniana, Ait.. 192
   Cerasus Caroliniana, Michx.
P. Chicasa, Michx.............. 192
   Cerasus Chicasa, Sering.
P. demissa, Walpers. 192
P. emarginata, Walpers.......... .. 193
   Var. *mollis,* Brewer.
   P. mollis. Walpers.
   Cerasus mollis, Dougl.
   C. glandulosus, Kellogg.
P. fasciculata. Gray............. 193
   Emplectocladus faciculatus, Torr.
P. illicifolia, Walpers........... 193
   Cerasus illicifollus, Nutt.
P. maritima. Wang............. 193
   P. littorallis. Bigelow.
   P. pygmœa, Willd.
P. Pennsylvanica 194
   Cerasus borealis, Michx.
   Cerasus Pennsylvanica, Sering.
P. pumila, Linn..... 195
   Cerasus pumila, Michx.
   C. glauca, Mœnch.
P. serotina, Ehr............. 195
   Cerasus Virginiana, Michx.
   Cerasus serotina, Loisel.
   P. Virginiana. Mill.
   P. cartilagenea, Lehm.
P. subcordata, Benth.............. 195
P. umbellata, Elliott............. 195
Pseudotsuga Douglasii, Carrière... 254
   P. Douglasii, Sabine.
   Abies Douglasii, Dougl.
   Tsuga Douglasii, Carrière.
   Var. *macrocarpa,* Engelm.
   Abies macrocarpa, Vasey.

Ptelea trifollata, Linn 284
Pterocarya fraxinifolia, Spach..... 283
  Juglans fraxinifolia, Lamk.
  J. pterocarya, Willd.
P. stenoptera, Cas. DC............. 283
  P. laevigata, Hort.
Quaking Asp................... 188
Quassia.................... 222
Qnercus, Linn 197
Q. agrifolia, Née... 197
  Q. oxyadenia, Torr.
Q. alba, Linn... 197
Q. aquatica, Catesby.............. 198
  Q. maritima, Willd.
Q. bicolor, Willd................. 198
  Q. Prinus, var. *tomentosa*, Michx.
  Q. Prinus, var. *discolor*, Michx.
  Q. Michauxii, Nutt.
  Var. *Michauxii*, Engelm.
  Q. Prinus palustris, Michx.
  Q. Michauxit, Nutt.
  Q. Prinus plantanoides, Lamk.
Q. Breweri, Engelm.............. 198
  Q. lobata, Engelm.
Q. Catesbæi, Michx... 199
Q. cerris, Linn.. 207
Q. cinerea, Michx. 199
  Q. Phellos, var. *cinerea*, Spach.
  Q. sempervirens, Catesby.
Q. chrysolepis, Liebm..... 199
  Q. fulvescens, Kellogg.
  Q. crassipocula, Torr.
  Var. *vaccinitfolia*, Engelm.
  Q. vacciniifolia, Kellogg.
Q. coccinea, Wang 199
  Q. ambigna, Michx.
  Q. borealis, Michx.
Q. densiflora, Hook. and Arn.... . 200
  Q. echinacea, Torr.
Q. Douglasii, Hook. and Arn...... 200
Q. dumosa, Nutt.. 200
  Q. berberidifolia, Liebm.
  Q. acutidens, Torr.
Q. Emoryi, Torr................. 201
  Q. hastata, Liebm.
Q. falcata, Michx............... 201
  Q. elongata, Willd.
  Q. discolor, var. *foltata*, Spach.
  Q. triloba, Michx.
  Q. falcata, var. *triloba*, DC.
Q. Garryana, Dougl.............. 201
  Q. Neæi, Liebm.
Q. Georgiana, M. A. Curtis........ 201
Q. heterophylla, Michx 201
  Q. aquatica, var. *heterophylla*, DC.
  Q. Phellos × *coccinea*, Engelm.
Q. hypoleuca, Engelm 202
  Q. conferlifolia, Torr.
Q. illicifolia, Wang............. 202
Q. imbricaria, Michx............ 202
Q. Kelloggii, Newberry......... 202
  Q. rubra, Benth.
  Q. tinctoria, var. *Californica*, Torr.
  Q. Sonomensis, Benth.
Q. laurifolia, Michx............ 202
  Q. aquatica, var. *laurifolia*, DC.
  Q. Phellos, var. *laurifolia*, Chap.
Q. lobata, Née..... 203

Q. Hindsii, Benth.
Q. Ransomi, Kellogg.
Q. lyrata, Walt.......... 203
Q. macrocarpa, Michx..... 203
  Q. olivæformis, Michx.
  Q. macrocarpa, var. *olivæformis*, Gray.
Q. Muhlenbergii, Engelm......... 203
  Q. castanea, Muhl. ap. Willd.
  Q. Prinus, var. *acuminata*, Michx.
Q. nigra, L. 204
  Q. ferruginea, Michx.
  Q. quinqueloba, Engelm.
  Q. nigra, var. *quinqueloba*, A.DC.
Q. oblougifolia, Torr............ 204
Q. Palmeri, Engelm... 204
  Q. chrysolepis, var. *Palmeri*, Englm.
Q. palustris, Du Roi... 204
  Q. rubra dissecta, Lamk.
Q. Phellos, Linn................. 204
Q. prinoides, Willd..... 205
  Q. Prinus pumila, Michx.
  Q. Chinquapin, Pursh.
Q. Prinus, L. 205
  Q. Prinus, var. *monticola*, Michx.
  Q. montana, Willd.
Q. Rober........... 207
Q. rubra, Linn.... 205
Q. stellata, Wang............... 205
  Q. obtusiloba, Michx.
  Q. Durandii, Buckley.
Q. tinctoria, Bartram... 205
  Q. nigra, Marsh.
  Q. velutina, Lam.
  Q. coccinea, var. *tinctoria*, Gray.
Q. tomentella, Engelm 206
Q. undulata, Torr............... 206
  Var. *Gambelii*, Engelm.
  Q. Gambelii, Nutt.
  Q. Drummondii, Liebm.
  Var. *Jamesii*, Engelm.
  Var. *Wrightii*, Engelm.
  Var. *breviloba*, Engelm.
  Q. obtusiloba, var. *breviloba*, Torr.
  Var. *oblongata*, Engelm.
  Q. oblongifolia, Torr.
  Var. *grisea*, Engelm.
  Q. grisea, Liebm.
  Var. *pungens*, Engelm.
  Q. pungens, Liebm.
Q. virens, Ait...... 206
  Q. sempervirens, Ait.
  Q. oleoides, Cham. and Schl.
  Q. retusa, Liebm.
  Q. maritima, Willd.
  Var. *maritima*, Chap.
Q. Wislizeni, A.DC.............. 207
  Q. Morehus, Kellogg.
Rabbit Berry.................... 221
Railroad Ties, Wood Used for..... 75
Raising Trees from Seed......... 25
Red Bay........................ 180
Red Bud........................ 131
Redwood................... 124, 247
Retinispora, Siebold 246
R. obtusa, Sieb 246
R. pisifera, Sieb................ 246
R. var. aurea plumosa............. 246

Rhamnus, Linn 208
R. alnifolia.... 208
R. Californin. Each... 208
   R. oleifolia. Hook.
   Frangula California, Gray.
   Var. tomentella 208
R. Caroliniana, Walt 208
   Frangula Caroliniana, Gray.
R. cathartices, Linn.............. 209
R. crocea, Nutt.................. 208
   R. illicifolia, Kellogg.
R. lanceolatus. Pursh............. 208
R. Purshiana, DC 208
   Frangula Purshiana, Cooper.
Rhizophora Mangle, Linn 209
Rhododendron, Linn 209
R. catawbiense, Michx............ 210
R. Lapponicum, Wahl... 210
R. maximum, Linn 209
Rhns, Linn............. 210
R. aromatica, Alt................. 212
   R. trilobata, Nutt.
R. copallina, Linn.... 212
R. Cotinus, Linn............. 214
R. Cotonides, Nutt 211
R. diversiloba, Torr. and Gray... . 210
   R. lobata, Hook.
R. glabra, Linn... 212
   R. Caroliniana, Mill.
   R. elegans, Alt.
   Var. *laciniata.*
R. integrifolia, Benth and Hook.. 212
   Styphonia integrifolia, Nutt.
   S. serrata, Nutt.
R. laurina, Nutt. 214
   Lithræa laurina, Walp.
R. Metopium, Linn................ 214
R. Osbecki, DC................... 214
   R. semialata, Murr.
   R. alata, Sav.
   R. Japonica, Hort.
R. pumila, Michx.:.. 212
R. Toxicodendron, Linn... 211
   Var. *quercifolium,* Michx.
   Var. *radicans,* Torr.
R. typhina, Linn 214
Robinia, Linn..... 215
R. hispida, Linn..... 215
   R. rosea, Loisel.
R. Pseudacacia, Linn.............. 215
R. viscosa, Vent................. 215
   R. glutinosa, Curtis.
Rose acacia 215
Rose Bay....................... 209
Salisburia adiantifolia............. 236
Salix, Tour..................... 215
S. cordata, Muhl........... 216
   S. lutea, Nutt.
   Var. *Mackenziana,* Hook.
   Var. *Watsoni,* Bebb.
S. lævigata, Bebb................. 216
   Var. *angustifolia,* Bebb.
   Var. *congesta,* Bebb.
S. lasiandra, Benth............... 216
   S. Hoffmanniana, Hook. & Arn.
   S. speciosa, Nutt.
   S. arguta, var. *lasiandra,* Anders.
   Var. *typea,* Bebb.

Var. *lancifolia,* Bebb.
   S. lancifolia, Anders.
   Var. *Fendleriana,* Bebb.
   S. pentandra, var. *caudata,* Nutt.
S. lasiolepis, Benth............. 217
   Var. *Bigelowii,* Bebb.
   Var. *fallax,* Bebb.
S. lucida, Muhl.................. 217
S. nigra, Marsh.... 217
   S. ambigua, Pursh.
   S. Houstoniana, Pursh.
   S Caroliniana, Michx.
   S. falcata, Pursh.
Sambucus, Tour.................. 218
S. Canadensis, Linn.............. 219
S. glauca, Nutt. 218
S. pubens, Michx... 219
S. racemosa, Linn. 219
Sapindus, Linn 219
S. marginatus, Willd............. 219
S. Saponaria, Linn.............. 219
Sassafras officinale, Nees......... 219
   Laurus Sassafras, Linn.
   Persea Sassafras, Spreng.
Satin Wood..................... 231
Schæfferia frutescens, Jacq........ 220
   S. completa, Swartz.
   S. buxifolia, Nutt.
Schœpfia arborescens, R. & S..... 220
Sciadopitys verticillata, Siebold &
   Zucc.... 281
Screw Bean... 189
Screw-pod Mesquit..... 190
Sea-side Grape... 185
Season for Transplanting..... 66
Sebastiania lucida, Muell.......... 220
   Gymnanthes lucida, Swartz.
   Excœcaria lucida, Swartz.
Seedling Black Walnut... 30
Seedlings of Coniferæ............ 58
Seedling Maple 19
Seedling Pine.............. 61
Sequoia, Endl 247
S. gigantea, Decaisne............. 247
   Wellingtonia gigantea, Lindl.
   Washingtonia Californica, T.
   S. Wellingtonia, Laws.
   Taxodium giganteum, Kell. & Behr.
   Taxodium Washingtonianum, Wins-
   low.
S. sempervirens, Endl............ 246
   Taxodium sempervirens, Lamb.
   Schuberlia sempervirens, Spach.
Service Tree.................... 105
Shad Bush...................... 105
Sheep Berry................ 230
Shepherdia, Nutt................ 221
S. argentea, Nutt.. 221
S. Canadensis, Nutt............. 221
S. rotundifolia, Parry........... . 221
Sideroxylon, Linn............... 222
S. mastichodendron, Jacq......... 222
   S. pallidum, Spreng.
   Bumelia pallida, Swartz.
   Bumelia fœtidissima, Nutt.
Silver Bell Tree................. 155
Simaruba, Aublet.............. ... 222
S. glauca, HBK................. 222

Snowball.......... 231
Snow-drop Tree.................... 155
Soap Berry...................... 219
Sophora, Linn...... 222
S. affinis, Torr. & Gray............ 222
　Styphnolobium affine, Walp.
S. Japonica, Sieb............... ... 223
S. secundiflora, Lag. 222
　S. speciosa, Torr.
S. tomentosa, Linn................ 223
Sorrel Tree... 178
Sour Gum. 175
Sowing Seeds........ 60
Sweet Gale...................... 174
Sweet Gum.. 164
Sweet Leaf...................... ... 224
Spanish Buckeye............... 229
Spice Tree................... 228
Spoonwood..... 163
Spruce, Douglass... 246
　Black or Double............. 256
　Engelmann's... 256
　Norway 257
　Silver 257
　Sitcha....................... 257
　Tiger's Tail............. 258
Stinking Cedar.. 235
Striped Dog-wood... 92
Stuartia, Catesby......... 223
S. Japonica, Sieb. & Zucc.... 224
S. pentagynia, L'Her............. 224
S. Virginica, Cav........ 224
Sugar-berry........................ 129
Sumach 210
　Coral........................ 214
　Dwarf 212
　Entire-leaved........ 212
　Evergreen 214
　Fragrant 212
　Oak-leaved............... .. 211
　Smooth...................... 212
　Staghorn................... 211
　Venetian.... 213
Swietenia Mahagoni, Linn........ 224
Sycamore........ 184
Symplocos tinctoria, L'Her....... 224
　Hopea tinctorea, Linn.
Tacamahac..................... 186
Tamarack........ 259
Taxodium distichum, Richard.. .. 249
　Cupressus disticha, Linn.
Taxus, Tour...................... 235
T. adpressa, Knight............... 234
T. baccata, Linn., var. Canadensis,
　Gray.. 233
　T. Canadensis, Willd.
T. brevifolia, Nutt............... . 234
　T. baccata, Hook.
　T. Boursieri, Carrière.
　T. Lindleyana, Muir.
T. cuspidata, Sieb 234
T. Floridana, Nutt......... 234
T. globosa, Schlect 234
Torreya, Arnott................. 234
T. Californica, Torr............ 234
　T. Myristica, Muir.
T. grandis, Fortune....... 236
T. nucifera, Zucc.................. 235

Torreya, Nut-bearing............. 235
　Tall....... 236
T. taxifolia, Arn..... 235
Thorn, Barberry-leaved............. 139
　Cockspur. 139
　Douglass... 139
　English Hawthorn.............. 141
　Evergreen................. 141
　Parsley-leaved.................. 139
　Pear or Black............... 140
　Scarlet-fruited............... 139
　Small-leaved................. 140
　Spatula-leaved................. 140
　Tall Hawthorn............... .. 139
　Yellow-fruited 139
Three-thorned Acacia... 152
Thuya, Tour........ 244
T. gigantea, Nutt............. ... 245
　T. plicata, Donn.
　T. Menziesii, Dougl.
T. occidentalis, Linn.............. 245
Tilia Americana, Linn............ 225
　Var. macrophylla, Hort.
　Var. pubescens, Gray.
　T. pubescens, Ait.
　T. laxiflora, Michx.
Tilia Europæa, Linn........ 226
T. heterophylla, Vent.............. 225
　T. alba, Michx.
　T. laxiflora, Pursh.
Time to Prune. 69
Tooth-ache Tree........ 231
Tornilla....................... 190
Torch Wood..................... 106
Toyon......................... 156
Transplanting Seedlings........ .. 32
Tree-Digger 29
Trees for Shelter............. 17
Tsuga Canadensis, Carrière. 254
　Pinus Canadensis, Linn.
　Abies Canadensis, Michx.
　Picea Canadensis, Link.
T. Mertensiana. Bongard.......... 255
　Pinus Mertensiana, Bong.
　Abies Mertensiana, Lindl.
　Abies Albertiana, Murr.
　Abies Bridgesii, Kellogg.
T. Pattoniana, Engelm........... 255
　Abies Pattoniana, Jeffrey.
　Abies Hookeriana, Murr.
　Abies Williamsonii, Newberry.
　Pinus Pattoniana, Parl.
Tulip Tree.................... 166
Tupelo... 175
　Large....................... 176
Ulmus, Linn.................... 226
U. alata, Michx 227
　U. pumila, Nutt.
U. Americana, Willd 227
　U. Floridana, Chapman.
U. crassifolia, Nutt............. . 227
　U. opaca, Nutt.
U. fulva, Michx...... 227
　U. rubra, Michx.
U. racemosa, Thomas............ 228
Umbellularia Californica, Nutt ... 229
　Oreodaphne Californica, Nees.
　Tetranthera Californica, H. & Arn.

Drimyphyllum panciflorum, Nutt.
Ungnadia speciosa, Endl 229
Umbrella Tree... 170
Viburnum, Linn.................. 230
V. Lentago. Linn................ 230
V. opulus, Linn..... 230
   Var. *oxycoccus*, Pursh.
   Var. *edule*, Pursh.
V. prunifolium. Linn 230
Virgilia lutea. Michx....... 123
Virginia Poplar.................. 166
Walnut....................... 158
   Barthere 161
   Black....... 159
   Californian................ 159
   Cut-leaved 162
   English 160
   French 160
   Gibbous.................... 161
   Small-fruited............ 161
Water Beech...................... 115
Wax Myrtle...................... 174
West Indian Birch.............. ... 114
Whahoo.... 227
White Basewood.. 225
White Fringe Tree..... 132
White Mangrove 164
White Wood 166

Wicky.................. 163
Willow, Babylonian......... 218
   Black...... 217
   Heart-leaved 216
   Hooped-leaved............. ... 218
   Long-leaved............... 216
   Ring-leaved. 218
   Shining... 217
   Weeping 218
   White... 218
Xanthoxylum, Linn............. 231
X. Americanum, Mill. 232
X. Caribæum.................. 231
   X. Floridanum, Nutt.
X. Clava-Herculis, Linn........... 231
   X. Carolinianum, Lam.
X. Pterota, HBK.. 231
Ximenia Americana, Linn........ 232
Yeara...................... 210
Yellow Cucumber Tree.... 168
Yellow Wood 133
Yew, American.............. 233
   Canada.................. 232
   Florida.................... 234
   Mexican.... 234
   Western................ 234
Zizyphus obtusifolius, Gray...... 232
   Paliurus Texensis, Scheele.

STANDARD BOOKS

ORANGE JUDD COMPANY

NEW YORK
439-441 Lafayette Street

CHICAGO
Marquette Building

---

*B*OOKS *sent to all parts of the world for catalog price. Discounts for large quantities on application. Correspondence invited. Brief descriptive catalog free. Large illustrated catalog, six cents.*

Soils

By CHARLES WILLIAM BURKETT, Director Kansas Agricultural Experiment Station. The most complete and popular work of the kind ever published. As a rule, a book of this sort is dry and uninteresting, but in this case it reads like a novel. The author has put into it his individuality. The story of the properties of the soils, their improvement and management, as well as a discussion of the problems of crop growing and crop feeding, make this book equally valuable to the farmer, student and teacher.

There are many illustrations of a practical character, each one suggesting some fundamental principle in soil management. 303 pages. 5½ x 8 inches. Cloth. $1.25

Insects Injurious to Vegetables

By Dr. F. H. CHITTENDEN, of the United States Department of Agriculture. A complete, practical work giving descriptions of the more important insects attacking vegetables of all kinds with simple and inexpensive remedies to check and destroy them, together with timely suggestions to prevent their recurrence. A ready reference book for truckers, market-gardeners, farmers as well as others who grow vegetables in a small way for home use; a valuable guide for college and experiment station workers, school-teachers and others interested in entomology of nature study. Profusely illustrated. 5½ x 8 inches. 300 pages. Cloth. $1.50

Alfalfa

By F. D. COBURN. Its growth, uses and feeding value. The fact that alfalfa thrives in almost any soil; that without reseeding it goes on yielding two, three, four and sometimes five cuttings annually for five, ten or perhaps 100 years; and that either green or cured it is one of the most nutritious forage plants known, makes reliable information upon its production and uses of unusual interest. Such information is given in this volume for every part of America, by the highest authority. Illustrated. 164 pages. 5 x 7 inches. Cloth. $0.50

Ginseng, Its Cultivation, Harvesting, Market' ing and Market Value

By MAURICE G. KAINS, with a short account of its history and botany. It discusses in a practical way how to begin with either seed or roots, soil, climate and location, preparation, planting and maintenance of the beds, artificial propagation, manures, enemies, selection for market and for improvement, preparation for sale, and the profits that may be expected. This booklet is concisely written, well and profusely illustrated, and should be in the hands of all who expect to grow this drug to supply the export trade, and to add a new and profitable industry to their farms and gardens without interfering with the regular work. New edition. Revised and enlarged. Illustrated. 5 x 7 inches. Cloth. . . . $0.50

Landscape Gardening

By F. A. WAUGH, professor of horticulture, University of Vermont. A treatise on the general principles governing outdoor art; with sundry suggestions for their application in the commoner problems of gardening. Every paragraph is short, terse and to the point, giving perfect clearness to the discussions at all points. In spite of the natural difficulty of presenting abstract principles the whole matter is made entirely plain even to the inexperienced reader. Illustrated. 152 pages. 5 x 7 inches. Cloth. $0.50

Hedges, Windbreaks, Shelters and Live Fences

By E. P. POWELL. A treatise on the planting, growth and management of hedge plants for country and suburban homes. It gives accurate directions concerning hedges; how to plant and how to treat them; and especially concerning windbreaks and shelters. It includes the whole art of making a delightful home, giving directions for nooks and balconies, for bird culture and for human comfort. Illustrated. 140 pages. 5 x 7 inches. Cloth. $0.50

Cabbage, Cauliflower and Allied Vegetables

By C. L. ALLEN. A practical treatise on the various types and varieties of cabbage, cauliflower, broccoli, Brussels sprouts, kale, collards and kohl-rabi. An explanation is given of the requirements, conditions, cultivation and general management pertaining to the entire cabbage group. After this each class is treated separately and in detail. The chapter on seed raising is probably the most authoritative treatise on this subject ever published. Insects and fungi attacking this class of vegetables are given due attention. Illustrated. 126 pages. 5 x 7 inches. Cloth.' $0.50

Asparagus

By F. M. HEXAMER. This is the first book published in America which is exclusively devoted to the raising of asparagus for home use as well as for market. It is a practical and reliable treatise on the saving of the seed, raising of the plants, selection and preparation of the soil, planting, cultivation, manuring, cutting, bunching, packing, marketing. canning and drying insect enemies, fungous diseases and every requirement to successful asparagus culture, special emphasis being given to the importance of asparagus as a farm and money crop. Illustrated. 174 pages. 5 x 7 inches. Cloth.   - $0.50

The New Onion Culture

By T. GREINER. Rewritten, greatly enlarged and brought up to date. A new method of growing onions of largest size and yield, on less land, than can be raised by the old plan. Thousands of farmers and gardeners and many experiment stations have given it practical trials which have proved a success. A complete guide in growing onions with the greatest profit, explaining the whys and wherefores. Illustrated. 5 x 7 inches. 140 pages. Cloth. $0.50

The New Rhubarb Culture

A complete guide to dark forcing and field culture. Part I—By J. E. MORSE, the well-known Michigan trucker and originator of the now famous and extremely profitable new methods of dark forcing and field culture. Part II—Compiled by G. B. FISKE. Other methods practiced by the most experienced market gardeners, greenhouse men and experimenters in all parts of America. Illustrated. 130 pages. 5 x 7 inches. Cloth. . . , , , , , ... $0.50

Bean Culture

By GLENN C. SEVEY, B.S. A practical treatise on the production and marketing of beans. It includes the manner of growth, soils and fertilizers adapted, best varieties, seed selection and breeding, planting, harvesting, insects and fungous pests, composition and feeding value; with a special chapter on markets by Albert W. Fulton. A practical book for the grower and student alike. Illustrated. 144 pages. 5 x 7 inches. Cloth. $0.50

Celery Culture

By W. R. BEATTIE. A practical guide for beginners and a standard reference of great interest to persons already engaged in celery growing. It contains many illustrations giving a clear conception of the practical side of celery culture. The work is complete in every detail, from sowing a few seeds in a window-box in the house for early plants, to the handling and marketing of celery in carload lots. Fully illustrated. 150 pages. 5 x 7 inches. Cloth. $0.50

Tomato Culture

By WILL W. TRACY. The author has rounded up in this book the most complete account of tomato culture in all its phases that has ever been gotten together. It is no second-hand work of reference, but a complete story of the practical experiences of the best posted expert on tomatoes in the world. No gardener or farmer can afford to be without the book. Whether grown for home use or commercial purposes, the reader has here suggestions and information nowhere else available. Illustrated. 150 pages. 5 x 7 inches. Cloth. $0.50

The Potato

By SAMUEL FRASER. This book is destined to rank as a standard work upon Potato Culture. While the practical side has been emphasized, the scientific part has not been neglected, and the information given is of value, both to the grower and the student. Taken all in all, it is the most complete, reliable and authoritative book on the potato ever published in America. Illustrated. 200 pages. 5 x 7 inches. Cloth. $0.75

Dwarf Fruit Trees

By F. A. WAUGH. This interesting book describes in detail the several varieties of dwarf fruit trees, their propagation, planting, pruning, care and general management. Where there is a limited amount of ground to be devoted to orchard purposes, and where quick results are desired, this book will meet with a warm welcome. Illustrated. 112 pages. 5 x 7 inches. Cloth. $0.50

The Cereals in America

By THOMAS F. HUNT, M.S., D.Agri., Professor of Agronomy, Cornell University. If you raise five acres of any kind of grain you cannot afford to be without this book. It is in every way the best book on the subject that has ever been written. It treats of the cultivation and improvement of every grain crop raised in America in a thoroughly practical and accurate manner. The subject-matter includes a comprehensive and succinct treatise of wheat, maize, oats, barley, rye, rice, sorghum (kafir corn) and buckwheat, as related particularly to American conditions. First-hand knowledge has been the policy of the author in his work, and every crop treated is presented in the light of individual study of the plant. If you have this book you have the latest and best that has been written upon the subject. Illustrated. 450 pages. 5½ x 8 inches. Cloth. $1.75

The Forage and Fiber Crops in America

By THOMAS F. HUNT. This book is exactly what its title indicates. It is indispensable to the farmer, student and teacher who wishes all the latest and most important information on the subject of forage and fiber crops. Like its famous companion, "The Cereals in America," by the same author, it treats of the cultivation and improvement of every one of the forage and fiber crops. With this book in hand, you have the latest and most up-to-date information available. Illustrated. 428 pages. 5½ x 8 inches. Cloth. $1.75

The Book of Alfalfa

History. Cultivation and Merits. Its Uses as a Forage and Fertilizer. The appearance of the Hon. F. D. COBURN's little book on Alfalfa a few years ago has been a profit revelation to thousands of farmers throughout the country, and the increasing demand for still more information on the subject has induced the author to prepare the present volume, which is by far the most authoritative, complete and valuable work on this forage crop published anywhere. It is printed on fine paper and illustrated with many full-page photographs that were taken with the especial view of their relation to the text. 336 pages. 6½ x 9 inches. Bound in cloth, with gold stamping. It is unquestionably the handsomest agricultural reference book that has ever been issued. Price, postpaid . . . $2.00

Clean Milk

By S. D. BELCHER, M.D. In this book the author sets forth practical methods for the exclusion of bacteria from milk, and how to prevent contamination of milk from the stable to the consumer. Illustrated. 5 x 7 inches. 146 pages. Cloth. $1.00

www.ingramcontent.com/pod-product-compliance
Lightning Source LLC
Chambersburg PA
CBHW020500270326
41926CB00008B/686